Astronomy and Astrophysics in the New Millennium

Astronomy and Astrophysics Survey Committee
Board on Physics and Astronomy–Space Studies Board
Commission on Physical Sciences, Mathematics, and Applications
National Research Council

NATIONAL ACADEMY PRESS
Washington, D.C.

NOTICE: The project that is the subject of this report was approved by the Governing Board of the National Research Council, whose members are drawn from the councils of the National Academy of Sciences, the National Academy of Engineering, and the Institute of Medicine. The members of the committee responsible for the report were chosen for their special competences and with regard for appropriate balance.

This project was supported by the National Aeronautics and Space Administration under Grant No. NAG5-6916, the National Science Foundation under Grant No. AST-9800149, and the Keck Foundation.

Front Cover: The image is a portion of the Hubble Deep Field, the deepest image ever taken of the universe. The most distant galaxies in the Hubble Deep Field emitted their light when the universe was less than 1 billion years old—in other words, when it was less than 6 percent of its present age. In this image, we can establish that the most distant and therefore earliest galaxies were quite different from those we study nearby. They were smaller and less regular, as if they were being built up from primordial clumps of gas. But we have still not seen the very first galaxies and stars that were created after the Big Bang. Seeing the very first galaxies is the primary goal of the Hubble Space Telescope's successor, the Next Generation Space Telescope. Courtesy of the Space Telescope Science Institute.

Library of Congress Card Number: 00-112257
International Standard Book Numbers:
0-309-07031-7 (paperback)
0-309-07312-X (hardcover)

Additional copies of this report are available from:

National Academy Press, 2101 Constitution Avenue, N.W., Lockbox 285, Washington, DC 20055; (800) 624-6242 or (202) 334-3313 (in the Washington metropolitan area); Internet <http://www.nap.edu>

Board on Physics and Astronomy, National Research Council, HA 562, 2101 Constitution Avenue, N.W., Washington, DC 20418
Internet <http://www.national-academies.org/bpa>

Printed in the United States of America

THE NATIONAL ACADEMIES

National Academy of Sciences
National Academy of Engineering
Institute of Medicine
National Research Council

The **National Academy of Sciences** is a private, nonprofit, self-perpetuating society of distinguished scholars engaged in scientific and engineering research, dedicated to the furtherance of science and technology and to their use for the general welfare. Upon the authority of the charter granted to it by the Congress in 1863, the Academy has a mandate that requires it to advise the federal government on scientific and technical matters. Dr. Bruce M. Alberts is president of the National Academy of Sciences.

The **National Academy of Engineering** was established in 1964, under the charter of the National Academy of Sciences, as a parallel organization of outstanding engineers. It is autonomous in its administration and in the selection of its members, sharing with the National Academy of Sciences the responsibility for advising the federal government. The National Academy of Engineering also sponsors engineering programs aimed at meeting national needs, encourages education and research, and recognizes the superior achievements of engineers. Dr. William A. Wulf is president of the National Academy of Engineering.

The **Institute of Medicine** was established in 1970 by the National Academy of Sciences to secure the services of eminent members of appropriate professions in the examination of policy matters pertaining to the health of the public. The Institute acts under the responsibility given to the National Academy of Sciences by its congressional charter to be an adviser to the federal government and, upon its own initiative, to identify issues of medical care, research, and education. Dr. Kenneth I. Shine is president of the Institute of Medicine.

The **National Research Council** was organized by the National Academy of Sciences in 1916 to associate the broad community of science and technology with the Academy's purposes of furthering knowledge and advising the federal government. Functioning in accordance with general policies determined by the Academy, the Council has become the principal operating agency of both the National Academy of Sciences and the National Academy of Engineering in providing services to the government, the public, and the scientific and engineering communities. The Council is administered jointly by both Academies and the Institute of Medicine. Dr. Bruce M. Alberts and Dr. William A. Wulf are chairman and vice chairman, respectively, of the National Research Council.

ASTRONOMY AND ASTROPHYSICS SURVEY COMMITTEE

CHRISTOPHER F. McKEE, University of California, Berkeley, *Co-chair*
JOSEPH H. TAYLOR, JR., Princeton University, *Co-chair*
DAVID J. HOLLENBACH, NASA Ames Research Center, *Executive Officer*
TODD BOROSON, National Optical Astronomy Observatories
WENDY FREEDMAN, Carnegie Observatories
DAVID C. JEWITT, University of Hawaii
STEVEN M. KAHN, Columbia University
JAMES M. MORAN, JR., Harvard-Smithsonian Center for Astrophysics
JERRY E. NELSON, University of California Observatories
R. BRUCE PARTRIDGE, Haverford College
MARCIA RIEKE, University of Arizona
ANNEILA I. SARGENT, California Institute of Technology
ALAN TITLE, Lockheed-Martin Space Technology Center
SCOTT TREMAINE, Princeton University
MICHAEL S. TURNER, University of Chicago

NATIONAL RESEARCH COUNCIL STAFF

DONALD C. SHAPERO, Board on Physics and Astronomy, Director
JOSEPH K. ALEXANDER, Space Studies Board, Director
ROBERT L. RIEMER, Senior Program Officer
JOEL R. PARRIOTT, Program Officer
GRACE WANG, Administrative Associate (1998-1999)
SÄRAH A. CHOUDHURY, Project Associate (1999-2000)
MICHAEL LU, Project Assistant (1998-2000)
NELSON QUIÑONES, Project Assistant (2000)

PANEL ON ASTRONOMY EDUCATION AND POLICY

ANDREA K. DUPREE, Harvard-Smithsonian Center for Astrophysics, *Chair*
R. BRUCE PARTRIDGE, Haverford College, *Vice Chair* (education)
ANNEILA I. SARGENT, California Institute of Technology, *Vice Chair* (policy)
FRANK BASH, McDonald Observatory, University of Texas
GREGORY BOTHUN, University of Oregon
SUZAN EDWARDS, Smith College
RICCARDO GIACCONI, Associated Universities, Inc.
PETER A. GILMAN, National Center for Atmospheric Research
MICHAEL HAUSER, Space Telescope Science Institute
BLAIR SAVAGE, University of Wisconsin
IRWIN SHAPIRO, Harvard-Smithsonian Center for Astrophysics
FRANK SHU, University of California, Berkeley
NEIL DE GRASSE TYSON, American Museum of Natural History

PANEL ON BENEFITS TO THE NATION FROM ASTRONOMY AND ASTROPHYSICS

STEPHEN E. STROM, National Optical Astronomy Observatories, *Chair*
DAVID J. HOLLENBACH, NASA Ames Research Center, *Vice Chair*

CONTRIBUTORS TO THE PANEL

ROGER ANGEL, Steward Observatory, University of Arizona
DOUGLAS DUNCAN, American Astronomical Society; University of Chicago
ANDREW FRAKNOI, Foothills College
PAUL GOLDSMITH, National Astronomy and Ionosphere Center, Cornell University
NEAL KATZ, University of Massachusetts, Amherst
EUGENE LEVY, University of Arizona
STEPHEN MARAN, NASA Goddard Space Flight Center
DAVID MORRISON, NASA Ames Research Center
LEIF ROBINSON, Sky Publishing Corporation

WILLIAM SMITH, Association of Universities for Research in Astronomy, Inc.
EDWARD STONE, California Institute of Technology, Jet Propulsion Laboratory
CHARLES TOWNES, University of California, Berkeley
VIRGINIA TRIMBLE, University of California, Irvine, and University of Maryland
PAUL VANDEN BOUT, National Radio Astronomy Observatory
SIDNEY WOLFF, National Optical Astronomy Observatories

PANEL ON HIGH-ENERGY ASTROPHYSICS FROM SPACE

ROGER D. BLANDFORD, California Institute of Technology, *Chair*
STEVEN M. KAHN, Columbia University, *Vice Chair*
LARS BILDSTEN, University of California, Berkeley
FRANCE A. CORDOVA, University of California, Santa Barbara
JONATHAN GRINDLAY, Harvard-Smithsonian Center for Astrophysics
DAN McCAMMON, University of Wisconsin
PETER MICHELSON, Stanford University
STEPHEN S. MURRAY, Harvard-Smithsonian Center for Astrophysics
RENE ASHWIN ONG, University of Chicago
CRAIG L. SARAZIN, University of Virginia
NICHOLAS WHITE, NASA Goddard Space Flight Center
STANFORD EARL WOOSLEY, University of California, Santa Cruz

PANEL ON OPTICAL AND INFRARED ASTRONOMY FROM THE GROUND

ALAN DRESSLER, Carnegie Observatories, *Chair*
TODD BOROSON, National Optical Astronomy Observatories, *Vice Chair*
JERRY E. NELSON, University of California Observatories, *Vice Chair*
JILL BECHTOLD, University of Arizona
RAYMOND CARLBERG, University of Toronto
BRUCE CARNEY, University of North Carolina
JAMES ELLIOT, Massachusetts Institute of Technology
RICHARD ELSTON, University of Florida
ANDREA MIA GHEZ, University of California, Los Angeles

CHARLES LADA, Harvard-Smithsonian Center for Astrophysics
JAMES W. LIEBERT, University of Arizona
CHARLES C. STEIDEL, California Institute of Technology
CHRISTOPHER STUBBS, University of Washington
DAVID C. JEWITT, University of Hawaii, *Ex Officio*

PANEL ON PARTICLE, NUCLEAR, AND GRAVITATIONAL-WAVE ASTROPHYSICS

THOMAS K. GAISSER, University of Delaware, *Chair*
MICHAEL S. TURNER, University of Chicago, *Vice Chair*
BARRY BARISH, California Institute of Technology
STEVEN WILLIAM BARWICK, University of California, Irvine
EUGENE BEIER, University of Pennsylvania
JOSHUA FRIEMAN, Fermi National Accelerator Laboratory
ALICE KUST HARDING, NASA Goddard Space Flight Center
RICHARD ALWIN MEWALDT, California Institute of Technology
RENE ASHWIN ONG, University of Chicago
BOHDAN PACZYNSKI, Princeton University Observatory
BERNARD SADOULET, University of California, Berkeley
PIERRE SOKOLSKY, University of Utah
RAINER WEISS, Massachusetts Institute of Technology

PANEL ON RADIO AND SUBMILLIMETER-WAVE ASTRONOMY

MARTHA P. HAYNES, Cornell University, *Chair*
JAMES M. MORAN, JR., Harvard-Smithsonian Center for Astrophysics, *Vice Chair*
GEOFFREY A. BLAKE, California Institute of Technology
DONALD B. CAMPBELL, Cornell University
JOHN E. CARLSTROM, University of Chicago
NEAL J. EVANS, University of Texas at Austin
JACQUELINE N. HEWITT, Massachusetts Institute of Technology
KENNETH I. KELLERMANN, National Radio Astronomy Observatory
ALAN P. MARSCHER, Boston University
STEVEN T. MYERS, University of Pennsylvania

MARK J. REID, Harvard-Smithsonian Center for Astrophysics
WILLIAM J. WELCH, University of California, Berkeley
DONALD BACKER, University of California, Berkeley, *Consultant*

PANEL ON SOLAR ASTRONOMY

MICHAEL KNOELKER, University Corporation for Atmospheric Research, *Chair*
ALAN TITLE, Lockheed-Martin Space Technology Center, *Vice Chair*
DALE EVERETT GARY, New Jersey Institute of Technology
PHILIP R. GOODE, New Jersey Institute of Technology
JOSEPH B. GURMAN, NASA Goddard Space Flight Center
SHADIA RIFAI HABBAL, Harvard-Smithsonian Center for Astrophysics
DANA WARFIELD LONGCOPE, Montana State University
RONALD LEE MOORE, NASA Marshall Space Flight Center
THOMAS RIMMELE, National Solar Observatory
JOHN H. THOMAS, University of Rochester
ELLEN GOULD ZWEIBEL, University of Colorado, Boulder

PANEL ON THEORY, COMPUTATION, AND DATA EXPLORATION

WILLIAM H. PRESS, Los Alamos National Laboratory, *Chair*
SCOTT TREMAINE, Princeton University, *Vice Chair*
CHARLES ALCOCK, Lawrence Livermore National Laboratory/University of Pennsylvania
LARS BILDSTEN, University of California, Berkeley/Santa Barbara
ADAM BURROWS, University of Arizona
LARS HERNQUIST, Harvard-Smithsonian Center for Astrophysics
CRAIG JAMES HOGAN, University of Washington
MARC PAUL KAMIONKOWSKI, Columbia University
MICHAEL NORMAN, University of Illinois at Urbana-Champaign
EVE OSTRIKER, University of Maryland
THOMAS A. PRINCE, California Institute of Technology
ALEX SANDOR SZALAY, Johns Hopkins University
ROBERT F. STEIN, Michigan State University, *Consultant*

PANEL ON ULTRAVIOLET, OPTICAL, AND INFRARED ASTRONOMY FROM SPACE

STEVEN V.W. BECKWITH, Space Telescope Science Institute, *Chair*
WENDY FREEDMAN, Carnegie Observatories, *Vice Chair*
MARCIA RIEKE, University of Arizona, *Vice Chair*
JOSEPH A. BURNS, Cornell University
DALE CRUIKSHANK, NASA Ames Research Center
RICHARD S. ELLIS, University of Cambridge
ALEXEI V. FILIPPENKO, University of California, Berkeley
MARTIN O. HARWIT, Washington, D.C.
LYNNE HILLENBRAND, California Institute of Technology
SHRINIVAS KULKARNI, California Institute of Technology
ABRAHAM LOEB, Harvard-Smithsonian Center for Astrophysics
ROBERT D. MATHIEU, University of Wisconsin
WARREN MOOS, Johns Hopkins University
J. MICHAEL SHULL, University of Colorado
EDWARD L. WRIGHT, University of California, Los Angeles
DAVID C. JEWITT, University of Hawaii, *Ex Officio*

AD HOC CROSS-PANEL WORKING GROUPS

Astronomical Surveys, Thomas A. Prince, *Chair*
Extrasolar Planets, David C. Jewitt, *Chair*
Laboratory Astrophysics, Charles Alcock, *Chair*
NSF-Funded National Observatories, Frank Bash, *Chair*

BOARD ON PHYSICS AND ASTRONOMY

SPACE STUDIES BOARD

COMMISSION ON PHYSICAL SCIENCES, MATHEMATICS, AND APPLICATIONS

Preface

In 1997, the Board on Physics and Astronomy asked BPA member Anthony Readhead and director Don Shapero to convene a small group of leading astronomers to consider the need for a new decadal survey of astronomy and astrophysics. The group concluded that the time was ripe for a new decadal survey in the 50-year series of such studies. It recommended the establishment of a new Astronomy and Astrophysics Survey Committee to carry out a broad scientific assessment of the field and to recommend new ground- and space-based programs for the decade 2000 to 2010. It also considered the framework for the survey, which ultimately led to the following detailed charge to the committee:

> The committee will survey the field of space- and ground-based astronomy and astrophysics, recommending priorities for the most important new initiatives of the decade 2000-2010. The principal goal of the study will be an assessment of proposed activities in astronomy and astrophysics and the preparation of a concise report addressed to the agencies supporting the field, the congressional committees with jurisdiction over these agencies, and the scientific community. The study will restrict its scope to experimental and theoretical aspects of subfields involving remote observations from the Earth and space and analysis of astronomical objects. Missions to make in situ studies of the Earth and planets, which have been treated by other National Research Council and Academy reports, will be excluded. Attention will be given to effective implementation of proposed and existing programs and to the organizational infrastructure and the human aspects of the field involving demography and education. Promising areas for the development of new technologies will be suggested.
>
> A brief review of the initiatives of other nations will be given together with a discussion of the possibilities of joint ventures and other forms of international cooperation. Prospects for combining resources—private, state, federal, and international—to build the strongest program possible

> for U.S. astronomy will be explored. Recommendations for new initiatives will be presented in priority order within different categories.
>
> The committee will also address two questions posed by the House Science Committee staff: Have NASA and NSF mission objectives resulted in a balanced, broad-based, robust science program for astronomy? That is, NASA's mission is to fund research that supports flight programs and focused campaigns such as Origins, whereas NSF's mission is to support basic research. Have these overall missions been adequately coordinated and has this resulted in an optimum science program from a productivity standpoint? What special strategies are needed for strategic cooperation between NASA and NSF? Should these be included in agency strategic plans? How do NASA and NSF determine the relative priority of new technological opportunities (including new facilities) compared to providing long-term support for associated research grants and facility operations?
>
> The committee will consult widely within the astronomical and astrophysical community and make a concerted effort to disseminate its recommendations promptly and effectively.

The two major questions posed by the House Science Committee staff (detailed above) were accompanied by several other questions that were treated in a report entitled *Federal Funding of Astronomical Research,* prepared by the Committee on Astronomy and Astrophysics (National Academy Press, Washington, D.C., 2000). That report was submitted to the survey committee as input to its deliberations.

The National Research Council established the survey under the auspices of the BPA, which oversaw the study in close consultation with the Space Studies Board. After consultations with members of the National Academy of Sciences Astronomy Section, members of astronomy departments in U.S. universities, and other leading astronomers, the BPA presented a slate of nominees for membership on the survey committee to the chair of the National Research Council. The NRC chair subsequently appointed the 15-member Astronomy and Astrophysics Survey Committee (AASC), with Joseph H. Taylor, Jr., and Christopher F. McKee as co-chairs, to carry out the study.

To provide detailed input to the AASC on the wavelength-based subdisciplines of astronomy and other areas, nine panels were established. Each panel's vice chair was selected from the membership of the AASC. The panel vice chairs were thus able to serve as liaisons between the panels and the main committee and to articulate the priorities of the subdisciplines within the AASC in the process of setting priorities. The panels included more than 100 people, who together were able to encompass the enormous intellectual breadth of modern astronomy and astrophysics.

Each panel met three times and also held two open "town meeting" sessions at the January and June 1999 meetings of the American Astronomical Society. Many of the panel members also held sessions at other professional gatherings, as well as at astronomical departments and centers throughout the United States.

The seven science panels were charged with preparing reports that identified the most important scientific goals in their respective areas, prioritizing the new initiatives needed to achieve these goals, recommending proposals for technology development, considering the possibilities for international collaboration, and discussing any policy issues relevant to their charge. The science panels were

- High-energy Astrophysics from Space;
- Optical and Infrared Astronomy from the Ground;
- Particle, Nuclear, and Gravitational-Wave Astrophysics;
- Radio and Submillimeter-Wave Astronomy;
- Solar Astronomy;
- Theory, Computation, and Data Exploration; and
- Ultraviolet, Optical, and Infrared Astronomy from Space.

Their reports are published in a separate volume entitled *Astronomy and Astrophysics in the New Millennium: Panel Reports* (National Academy Press, Washington, D.C., 2001).

The reports of the other two panels—Astronomy Education and Policy, and Benefits to the Nation from Astronomy and Astrophysics—were revised and incorporated into the AASC main report. As mentioned above, the AASC also drew on the report *Federal Funding of Astronomical Research* as well as other NRC reports cited in the text. Further valuable input to the AASC and its panels was provided by four ad hoc cross-panel working groups: Astronomical Surveys (T. Prince, Chair), Extrasolar Planets (D. Jewitt, Chair), Laboratory Astrophysics (C. Alcock, Chair), and NSF-Funded National Observatories (F. Bash, Chair).

Members of the survey committee and the panels consulted widely with their colleagues to solicit advice and to inform other members of the astronomical community of the main issues facing the committee. This consultation process provided useful input for the panel reports and also gave the survey committee a good picture of the community consensus on the various initiatives under consideration for inclusion among the priorities of the main report.

At the final AASC meeting in late 1999, the panel chairs participated with members of the survey committee to develop the new decadal survey's recommendations. The committee based its final recommendations and priorities in significant part on the panel reports and on the discussions with the panel chairs. As mentioned above, the panel reports, reviewed by the National Research Council together with the main report, are published in a separate volume subtitled *Panel Reports*. The overall priorities are presented in the present volume. The panel reports contain, in addition to more detailed discussion of these priorities, further projects and topics that were not selected by the AASC for inclusion among the overall priorities that are viewed as having importance for the field as a whole.

The AASC is grateful to the many astronomers, both in the United States and from abroad, who provided written advice or participated in organized discussions. We thank the National Science Foundation, the National Aeronautics and Space Administration, and the Keck Foundation for providing support for the project. We are grateful to Robert Milkey and Kevin Marvel and to the American Astronomical Society for assistance in the community outreach and town meeting sessions. The committee also acknowledges the assistance of NRC staff members, particularly the outstanding work of Joel Parriott and Roc Riemer, who provided support for the entire project, Susan Maurizi and Liz Fikre, who edited the reports, and the National Academy Press, which published the reports. We are also indebted to Robert Sokol and Ken Van Pool of Design@Large for their innovative design of the booklet that gives an overview of and popularizes the results of the survey. The timely completion of this report would not have been possible without the unstinting efforts of David Hollenbach, who served both as a member of the committee and as Executive Officer. Many other people too numerous to cite individually assisted in various aspects of the survey. We thank them all for their assistance.

Christopher F. McKee and Joseph H. Taylor, Jr., *Co-chairs*
Astronomy and Astrophysics Survey Committee

Acknowledgment of Reviewers

This report has been reviewed in draft form by individuals chosen for their diverse perspectives and technical expertise, in accordance with procedures approved by the NRC's Report Review Committee. The purpose of this independent review is to provide candid and critical comments that will assist the institution in making its published report as sound as possible and to ensure that the report meets institutional standards for objectivity, evidence, and responsiveness to the study charge. The review comments and draft manuscript remain confidential to protect the integrity of the deliberative process. We wish to thank the following individuals for their review of this report and/or one or more of the panel reports:

W. David Arnett, Steward Observatory, University of Arizona,
Peter Banks, ERIM International, Inc. (retired),
Gordon A. Baym, University of Illinois at Urbana-Champaign,
Roger Chevalier, University of Virginia,
Anita L. Cochran, University of Texas at Austin,
Marshall H. Cohen, California Institute of Technology,
Anne P. Cowley, Arizona State University,
Val L. Fitch, Princeton University,
Bill Green, former Congressman, New York,
Karen L. Harvey, Solar Physics Research Group,
John P. Huchra, Harvard-Smithsonian Center for Astrophysics,
Robert P. Kirshner, Harvard-Smithsonian Center for Astrophysics,
Chryssa Kouveliotou, NASA Marshall Space Flight Center,
Richard G. Kron, Yerkes Observatory,
Jeffrey Linsky, University of Colorado/JILA,

Richard McCray, University of Colorado/JILA,
Melissa McGrath, Space Telescope Science Institute,
Mark Morris, University of California, Los Angeles,
Martin J. Rees, Institute of Astronomy, Cambridge University, U.K.,
Morton S. Roberts, National Radio Astronomy Observatory–Charlottesville,
Patrick Thaddeus, Harvard-Smithsonian Center for Astrophysics,
J. Anthony Tyson, Lucent Technologies, and
David T. Wilkinson, Princeton University.

Although the reviewers listed above have provided many constructive comments and suggestions, they were not asked to endorse the conclusions or recommendations, nor did they see the final draft of the report before its release. The review of this report and of the panel reports was overseen by Nicholas P. Samios, Brookhaven National Laboratory, appointed by the NRC's Commission on Physical Sciences, Mathematics, and Applications, and Lewis M. Branscomb, John F. Kennedy School of Government, Harvard University, appointed by the Report Review Committee, who were responsible for making certain that an independent examination of the reports was carried out in accordance with institutional procedures and that all review comments were carefully considered. Responsibility for the final content of this report and the panel reports rests entirely with the authoring committee and the institution.

Contents

Executive Summary

ASTRONOMY AND ASTROPHYSICS IN THE NEW MILLENNIUM

In the first decade of the new millennium, we are poised to take a giant step forward in understanding the universe and our place within it. The decade of the 1990s saw an enormous number of exciting discoveries in astronomy and astrophysics. For example, humanity's centuries-long quest for evidence of the existence of planets around other stars resulted in the discovery of extrasolar planets, and the number of planets known continues to grow. Astronomers peered far back in time, to only a few hundred thousand years after the Big Bang, and found the seeds from which all galaxies, such as our own Milky Way, were formed. At the end of the decade came evidence for a new form of energy that may pervade the universe. Nearby galaxies were found to harbor extremely massive black holes in their centers. Distant galaxies were discovered near the edge of the visible universe. In our own solar system, the discovery of Kuiper Belt objects—some of which lie beyond the orbit of Pluto—opens a new window onto the history of the solar system. This report presents a comprehensive and prioritized plan for the new decade that builds on these and other discoveries to pursue the goal of understanding the universe, a goal that unites astronomers and astrophysicists with scientists from many other disciplines.

The Astronomy and Astrophysics Survey Committee was charged with surveying both ground- and space-based astronomy and recommending priorities for new initiatives in the decade 2000 to 2010. In addition, the committee was asked to consider the effective implementation of both the proposed initiatives and the existing programs. The committee's charge excludes in situ studies of Earth and the planets, which are covered by other National Research Council committees: the Committee on Planetary and Lunar Exploration and the Committee on Solar and Space Physics. To carry out its mandate, the committee established nine panels with more than 100 distinguished members of the astronomical community. Broad input was sought through the panels, in forums held by the American Astronomical Society, and in meetings with representatives of the international astronomical community. The committee's recommendations build on those of four previous decadal surveys (NRC, 1964, 1972, 1982, 1991), in particular the report of the 1991 Astronomy and Astrophysics Survey Committee, *The Decade of Discovery in Astronomy and Astrophysics* (referred to in this report as the 1991 survey; also known as the Bahcall report).

The fundamental goal of astronomy and astrophysics is to understand how the universe and its constituent galaxies, stars, and planets formed, how they evolved, and what their destiny will be. To achieve this goal, researchers must pursue a strategy with several elements:

- Survey the universe and its constituents, including galaxies as they evolve through cosmic time, stars and planets as they form out of collapsing interstellar clouds in our galaxy, interstellar and intergalactic gas as it accumulates the elements created in stars and supernovae, and the mysterious dark matter and perhaps dark energy that so strongly influence the large-scale structure and dynamics of the universe.
- Use the universe as a unique laboratory for probing the laws of physics in regimes not accessible on Earth, such as the very early universe or near the event horizon of a black hole.
- Search for life beyond Earth, and if it is found, determine its nature and its distribution.
- Develop a conceptual framework that accounts for all that astronomers have observed.

Several key problems are particularly ripe for advances in this decade:

- Determine the large-scale properties of the universe: the amount, distribution, and nature of its matter and energy, its age, and the history of its expansion.
- Study the dawn of the modern universe, when the first stars and galaxies formed.
- Understand the formation and evolution of black holes of all sizes.
- Study the formation of stars and their planetary systems, and the birth and evolution of giant and terrestrial planets.
- Understand how the astronomical environment affects Earth.

These scientific themes, all of which now appear to offer particular promise for immediate progress, are only part of the much larger tapestry that is modern astronomy and astrophysics. For example, scientists cannot hope to understand the formation of black holes without understanding the late stages of stellar evolution, and the full significance of observations of the galaxies in the very early universe will not be clear until it is clear how these galaxies have evolved since that time. Although the new initiatives that the committee recommends will advance knowledge in many other areas as well, they were selected explicitly to address one or more of the important themes listed above.

In addition, the committee believes that astronomers can make important contributions to education. Building on widespread interest in astronomical discoveries, astronomers should:

- Use astronomy as a gateway to enhance the public's understanding of science and as a catalyst to improve teachers' education in science and to advance interdisciplinary training of the technical work force.

OPTIMIZING THE RETURN ON THE NATION'S INVESTMENT IN ASTRONOMY AND ASTROPHYSICS

The United States has been generous in its support of astronomy and astrophysics and as a result enjoys a leading role in almost all areas of astronomy and astrophysics. So that the nation can continue to obtain maximum scientific return on its investment, the committee makes several recommendations to optimize the system of support for astronomical research.

BALANCING NEW INITIATIVES WITH THE ONGOING PROGRAM

An effective program of astronomy and astrophysics research must balance the need for initiatives to address new opportunites with completion of projects accorded high scientific priority in previous surveys.

- **The committee reaffirms the recommendations of the 1991 Astronomy and Astrophysics Survey Committee (NRC, 1991) by endorsing the completion of the Space Infrared Telescope Facility (SIRTF), the Millimeter Array (MMA; now part of the Atacama Large Millimeter Array, or ALMA), the Stratospheric Observatory for Infrared Astronomy (SOFIA), and the Astrometric Interferometry Mission (now called the Space Interferometry Mission, or SIM). Consistent with the recommendations of the Task Group on Space Astronomy and Astrophysics (NRC, 1997), the committee stresses the importance of studying the cosmic**

microwave background with the Microwave Anisotropy Probe (MAP) mission, the European Planck Surveyor mission, and ground-based and balloon programs.

The committee endorses U.S. participation in the European Far Infrared Space Telescope (FIRST), and it endorses the planned continuation of the operation of the Hubble Space Telescope (HST) at a reduced cost until the end of the decade.

- **To achieve the full scientific potential of a new facility, it is essential that, prior to construction, funds be identified for operation of the facility, for renewal of its instrumentation, and for grants for data analysis and the development of associated theory.**

NASA already follows this recommendation in large part by including Mission Operations and Data Analysis (MO&DA) in its budgeting for new missions. The committee recommends that funds for associated theory be included in MO&DA as well. It recommends further that the National Science Foundation include funds for facility operation, renewal of instrumentation, and grants for data analysis and theory along with the construction costs in the budgets for all new federally funded, ground-based facilities. These recommendations are consistent with those of the 1991 survey. For the purpose of total project budget estimation, the committee adopted a model in which operation amounts to 7 percent of the capital cost per year and instrumentation amounts to 3 percent per year for the first 5 years of operation. The committee recommends that total project budgets provide for grants for data analysis and associated theory at the rate of 3 percent of the capital cost per year for major facilities and 5 percent per year for moderate ones. On the basis of this model, the committee has included funds for operations, instrumentation, and grants for a period of 5 years in the cost estimates provided in this report for most ground-based initiatives.

- **Adequate funding for unrestricted grants that provide broad support for research, students, and postdoctoral associates is required to ensure the future vitality of the field; therefore new initiatives should not be undertaken at the expense of the unrestricted grants program.**

Grants not tied to a facility or program—unrestricted grants—often drive the future directions of astronomy.

STRENGTHENING GROUND-BASED ASTRONOMY AND ASTROPHYSICS

The committee addresses several structural issues in ground-based astronomy and astrophysics.

- **U.S. ground-based optical and infrared facilities, radio facilities, and solar facilities should each be viewed by the National Science Foundation (NSF) and the astronomical community as a single integrated system drawing on both federal and nonfederal funding sources. Effective national organizations are essential to coordinate, and to ensure the success and efficiency of, these systems. Universities and independent observatories should work with the national organizations to ensure the success of these systems.**
- **Cross-disciplinary competitive reviews should be held about every 5 years for all NSF astronomy facilities. In these reviews, it should be standard policy to set priorities and consider possible closure or privatization.**

The National Radio Astronomy Observatory (NRAO) and the National Astronomy and Ionosphere Center (NAIC) currently serve as effective national organizations for radio astronomy, and the National Solar Observatory (NSO) does so for solar physics. The National Optical Astronomy Observatories (NOAO) as currently functioning and overseen does not fulfill this role for ground-based optical and infrared astronomy. A plan for the transition of NOAO to an effective national organization for ground-based optical and infrared astronomy should be developed, and a high-level external review, based on appropriate, explicit criteria, should be initiated.

The Department of Energy (DOE) supports a broad range of programs in particle and nuclear astrophysics and in cosmology. The scientific payoff of this effort would be even stronger with a clearly articulated strategic plan for DOE's programs that involve astrophysics.

- **Given the increasing involvement of the Department of Energy in projects that involve astrophysics, the committee recommends that DOE develop a strategic plan for astrophysics that would lend programmatic coherence and facilitate coordination and cooperation with other agencies on science of mutual interest.**

ENSURING THE DIVERSITY OF NASA MISSIONS

NASA's Great Observatories have revolutionized understanding of the cosmos, while the extremely successful Explorer program provides targeted small-mission opportunities for advances in many areas of astronomy and astrophysics. The committee endorses the continuation of a vigorous Explorer program. There are now fewer opportunities for missions of moderate size, however, despite the enormous role such missions have played in the past.

- **NASA should continue to encourage the development of a diverse range of mission sizes, including small, moderate, and major, to ensure the most effective returns from the U.S. space program.**

INTEGRATING THEORY CHALLENGES INTO THE NEW INITIATIVES

The new initiatives recommended below are motivated in large part by theory, which is also key to interpreting the results. Adequate support for theory, including numerical simulation, is a cost-effective means for maximizing the impact of the nation's capital investment in science facilities. The committee therefore recommends that

- **To encourage theorists to contribute to the planning of missions and facilities and to the interpretation and understanding of the results, one or more explicitly funded theory challenges should be integrated with most moderate or major new initiatives.**

COORDINATING PROGRAMS AMONG FEDERAL AGENCIES

Because of the enormous scale of contemporary astronomical projects and the need for investigations that cross wavelength and discipline boundaries, cooperation among the federal agencies that support astronomical research often has benefits. To determine when interagency collaboration would be fruitful, each agency should have in place a strategic plan for astronomy and astrophysics and should also have cross-disciplinary committees (such as DOE and NSF's Scientific Assessment Group for Experiments in Non-Accelerator Physics

[SAGENAP] and NASA's Space Science Advisory Committee [SSAC]) available to evaluate proposed collaborations. The Office of Science and Technology Policy could play a useful role in facilitating such interagency cooperation.

COLLABORATING WITH INTERNATIONAL PARTNERS

International collaboration enables projects that are too costly for the United States alone and enhances the scientific return on projects by bringing in the scientific and technical expertise of international partners. In many cases, international collaboration provides opportunities for U.S. astronomers to participate in major international projects for a fraction of the total cost, as in the case of the European Solar and Heliospheric Observatory (SOHO), XMM-Newton, Planck Surveyor, and FIRST missions, and the Japanese Advanced Satellite for Cosmology and Astrophysics mission. Valuable opportunities for international collaboration exist for smaller missions as well. Collaborations on major projects require the full support of the participating scientific communities, which can be ensured if the projects are among the very highest priorities of the participants, as is the case with ALMA.

The committee affirms the value of international collaboration for ground- and space-based projects of all sizes. International collaboration plays a crucial role in a number of this committee's recommended initiatives, including the Next Generation Space Telescope, the Expanded Very Large Array, the Gamma-ray Large Area Space Telescope, the Laser Interferometer Space Antenna, the Advanced Solar Telescope, and the Square Kilometer Array technology development, and it could play a significant role in other recommended initiatives as well.

NEW INVESTMENTS IN ASTRONOMY AND ASTROPHYSICS

Many mysteries confront us in the quest to understand our place in the universe. How did the universe begin? What is the nature of the dark matter and the dark energy that pervade the universe? How did the first stars and galaxies form? Researchers infer the existence of stellar mass black holes in our galaxy and supermassive ones in the nuclei of galaxies. How did they form? The discovery of extrasolar planets has opened an entirely new chapter in astronomy, bringing a host of unresolved questions. How do planetary systems form and evolve? Are planetary systems like our solar system common in the universe? Do any extrasolar planetary systems harbor life? Even a familiar object like the Sun poses many mysteries. What causes the small variations in the Sun's luminosity that can affect Earth's climate? What is the origin of the eruptions on the solar surface that cause "space weather"?

To seek the answers to these questions and many others described in this report, the committee recommends a set of new initiatives for this decade that will substantially advance the frontiers of human knowledge. Table ES.1 presents these initiatives, combined for both ground- and space-based astronomy, in order of priority. The committee set the priorities primarily on the basis of scientific merit, but it also considered technical readiness, cost-effectiveness, impact on education and public outreach, and the relation to other projects. The initiatives were divided into three categories—major, moderate, and small—that were defined separately for ground- and space-based projects based on estimated cost (see Chapter 1). The estimated cost of the recommended program for the decade 2000 to 2010 is $4.7 billion in FY2000 dollars, about 20 percent greater than the $3.9 billion inflation-adjusted cost of the recommendations of the 1991 survey. Two of the recommended projects, the Terrestrial Planet Finder (TPF) and the Single Aperture Far Infrared (SAFIR) Observatory, could start near the end of this decade or at the beginning of the next. The committee has assumed that about 15 percent of the total estimated cost for these two projects will fall in this decade.

MAJOR INITIATIVES

- The **Next Generation Space Telescope (NGST),** the committee's top-priority recommendation, is designed to detect light from the first stars and to trace the evolution of galaxies from their formation to the present. It will revolutionize understanding of how stars and planets form in our galaxy today. NGST is an 8-m-class infrared space telescope with 100 times the sensitivity and 10 times the image sharpness of the Hubble Space Telescope in the infrared. Having NGST's sensitivity extend to 27 μm would add significantly to its scientific return. Technology development for this program is well under way. The European Space Agency and the Canadian Space Agency plan to make substantial contributions to the instrumentation for NGST.

TABLE ES.1 Prioritized Initiatives (Combined Ground and Space) and Estimated Federal Costs for the Decade 2000 to 2010[a,b]

Initiative	Cost[c] ($M)
Major Initiatives	
Next Generation Space Telescope (NGST)[d]	1,000
Giant Segmented Mirror Telescope (GSMT)[d]	350
Constellation-X Observatory (Con-X)	800
Expanded Very Large Array (EVLA)[d]	140
Large-aperture Synoptic Survey Telescope (LSST)	170
Terrestrial Planet Finder (TPF)[e]	200
Single Aperture Far Infrared (SAFIR) Observatory[e]	100
Subtotal for major initiatives	2,760
Moderate Initiatives	
Telescope System Instrumentation Program (TSIP)	50
Gamma-ray Large Area Space Telescope (GLAST)[d]	300
Laser Interferometer Space Antenna (LISA)[d]	250
Advanced Solar Telescope (AST)[d]	60
Square Kilometer Array (SKA) technology development	22
Solar Dynamics Observatory (SDO)	300
Combined Array for Research in Millimeter-wave Astronomy (CARMA)[d]	11
Energetic X-ray Imaging Survey Telescope (EXIST)	150
Very Energetic Radiation Imaging Telescope Array System (VERITAS)	35
Advanced Radio Interferometry between Space and Earth (ARISE)	350
Frequency Agile Solar Radio telescope (FASR)	26
South Pole Submillimeter-wave Telescope (SPST)	50
Subtotal for moderate initiatives	1,604
Small Initiatives	
National Virtual Observatory (NVO)	60
Other small initiatives[f]	246
Subtotal for small initiatives	306
DECADE TOTAL	4,670

[a]Cost estimates for ground-based capital projects include technology development plus funds for operations, new instrumentation, and facility grants for 5 years.

[b]Cost estimates for space-based projects exclude technology development.

[c]Best available estimated costs to U.S. government agencies in millions of FY2000 dollars and rounded. Full costs are given for all initiatives except TPF and the SAFIR Observatory.

[d]Cost estimate for this initiative assumes significant additional funding to be provided by international or private partner; see *Astronomy and Astrophysics in the New Millennium: Panel Reports* (NRC, 2001) for details.

[e]These missions could start at the turn of the decade. The committee attributes $200 million of the $1,700 million total estimated cost of TPF to the current decade and $100 million of the $600 million total estimated cost of the SAFIR Observatory to the current decade.

[f]See Chapter 1 for details.

- The **Giant Segmented Mirror Telescope (GSMT),** the committee's top ground-based recommendation and second priority overall, is a 30-m-class ground-based telescope that will be a powerful complement to NGST in tracing the evolution of galaxies and the formation of stars and planets. It will have unique capabilities in studying the evolution of the intergalactic medium and the history of star formation in our galaxy and its nearest neighbors. GSMT will use adaptive optics to achieve diffraction-limited imaging in the atmospheric windows between 1 and 25 μm and unprecedented light-gathering power between 0.3 and 1 μm. The committee recommends that the technology development for GSMT begin immediately and that construction start within the decade. Half the total cost should come from private and/or international partners. Open access to GSMT by the U.S. astronomical community should be directly proportional to the investment by the NSF.
- The **Constellation-X Observatory** is a suite of four powerful x-ray telescopes in space that will become the premier instrument for studying the formation and evolution of black holes of all sizes. Each telescope will have high spectral resolution over a broad energy range, enabling it to study quasars near the edge of the visible universe and to trace the evolution of the chemical elements. The technology issues are well in hand for a start in the middle of this decade.
- The **Expanded Very Large Array (EVLA)**—the revitalization of the VLA, the world's foremost centimeter-wave radio telescope—will take advantage of modern technology to attain unprecedented image quality with 10 times the sensitivity and 1,000 times the spectroscopic capability of the existing VLA. The addition of eight new antennas will provide an order-of-magnitude increase in angular resolution. With resolution comparable to that of ALMA and NGST, but operating at much longer wavelengths, the EVLA will be a powerful complement to these instruments for studying the formation of protoplanetary disks and the earliest stages of galaxy formation.
- The **Large-aperture Synoptic Survey Telescope (LSST)** is a 6.5-m-class optical telescope designed to survey the visible sky every week down to a much fainter level than that reached by

existing surveys. It will catalog 90 percent of the near-Earth objects larger than 300 m and assess the threat they pose to life on Earth. It will find some 10,000 primitive objects in the Kuiper Belt, which contains a fossil record of the formation of the solar system. It will also contribute to the study of the structure of the universe by observing thousands of supernovae, both nearby and at large redshift, and by measuring the distribution of dark matter through gravitational lensing. All the data will be available through the National Virtual Observatory (see below under "Small Initiatives"), providing access for astronomers and the public to very deep images of the changing night sky.

- The **Terrestrial Planet Finder (TPF)** is the most ambitious science mission ever attempted by NASA. It is currently envisaged as a free-flying infrared interferometer designed to study terrestrial planets around nearby stars—to find them, characterize their atmospheres, and search for evidence of life—and to obtain images of star-forming regions and distant galaxies with unprecedented resolution. The committee's recommendation of this mission is predicated on the assumptions that TPF will revolutionize major areas of both planetary and nonplanetary science and that, prior to the start of TPF, ground- and space-based searches will confirm the expectation that terrestrial planets are common around solar-type stars. Both NGST and SIM lie on the technology path necessary to achieve TPF.
- The **Single Aperture Far Infrared (SAFIR) Observatory** is an 8-m-class space-based telescope that will study the important and relatively unexplored spectral region between 30 and 300 μm. It will enable the study of galaxy formation and the earliest stage of star formation by revealing regions too enshrouded by dust to be studied by NGST, and too warm to be studied effectively with ALMA. As a follow-on to NGST, SAFIR could start toward the end of the decade, and it could form the basis for developing a far-infrared interferometer in the succeeding decade.

MODERATE INITIATIVES

GROUND-BASED PROGRAMS

The committee's recommended highest-priority moderate initiative overall is the Telescope System Instrumentation Program (TSIP), which would substantially increase NSF funding for instrumentation at large telescopes owned by independent observatories and provide new observing opportunities for the entire U.S. astronomical community. Its second priority among ground-based initiatives is the Advanced Solar Telescope (AST), which offers the prospect of revolutionizing understanding of magnetic phenomena in the Sun and in the rest of the universe. The committee's next recommendation is that a program be established to plan and develop technology for the Square Kilometer Array, an international centimeter-wave radio telescope for the second decade of the century. In order of priority, the other recommended moderate initiatives are the following: The Combined Array for Research in Millimeter-wave Astronomy (CARMA) will be a powerful millimeter-wave array in the Northern Hemisphere. The study of very-high-energy gamma rays will take a major step forward with the construction of the Very Energetic Radiation Imaging Telescope Array System (VERITAS). The Frequency Agile Solar Radio telescope (FASR) will apply modern technology to provide unique data on the Sun at radio wavelengths. The South Pole Submillimeter-wave Telescope (SPST) will take advantage of the extremely low opacity of the Antarctic atmosphere to carry out surveys at submillimeter wavelengths that are possible nowhere else on Earth.

Space-Based Programs

The committee's top recommendation for a moderate space-based mission is the Gamma-ray Large Area Space Telescope (GLAST). This joint NASA-DOE mission will provide observations of gamma rays from 10 MeV to 300 GeV with six times the effective area, six times the field of view, and substantially better angular resolution than the Energetic Gamma Ray Experiment aboard the Compton Gamma Ray Observatory. The committee's second-priority space-based project is the Laser Interferometer Space Antenna (LISA), which will be able to detect gravity waves from merging supermassive black holes throughout the visible universe and from close binary stars throughout our galaxy. The committee has assumed that LISA's cost will be shared with the European Space Agency. Four additional space-based missions have priority. The Solar Dynamics Observer (SDO), a successor to the pathbreaking SOHO mission, will study the outer convective zone of the Sun and the structure of the solar corona. The highly variable hard-x-ray sky will be mapped by the Energetic X-ray Imaging Survey Telescope (EXIST), which will be attached to the International Space Station. The Advanced Radio

Interferometry between Space and Earth (ARISE) mission is an orbiting antenna that will combine with the ground-based VLBA to provide an order-of-magnitude increase in resolution for studying the regions near supermassive black holes in active galactic nuclei.

SMALL INITIATIVES

Several small initiatives recommended by the committee span both ground and space. The first among them—the National Virtual Observatory (NVO)—is the committee's top priority among the small initiatives. The NVO will provide a "virtual sky" based on the enormous data sets being created now and the even larger ones proposed for the future. It will enable a new mode of research for professional astronomers and will provide to the public an unparalleled opportunity for education and discovery.

The remaining recommendations for small initiatives are not prioritized. The committee recommends establishing a laboratory astrophysics program and a national astrophysical theory postdoctoral program for both ground- and space-based endeavors. Augmentation of NASA's Astrophysics Theory Program will help restore a balance between the acquisition of data and the theory needed to interpret it. Ultralong-duration balloon flights offer the prospect of carrying out small space-based experiments at a small fraction of the cost of satellites. The Low Frequency Array (LOFAR), a joint Dutch-U.S. initiative, will dramatically increase knowledge of the universe at radio wavelengths longer than 2 m. The Advanced Cosmic-ray Composition Experiment for the Space Station (ACCESS) will address fundamental questions about the origin of cosmic rays. Expansion of the Synoptic Optical Long-term Investigation of the Sun (SOLIS) will permit investigation of the solar magnetic field over an entire solar cycle.

TECHNOLOGY

Technological innovation has often enabled astronomical discovery. Advances in technology in this decade are a prerequisite for many of the initiatives recommended in this report as well as for initiatives in the next decade. For the recommended space-based initiatives, technology investment as specified in the existing NASA technology road map is an assumed prerequisite for the cost estimates given in Table ES.1. It is essential to maintain funding for these initiatives if NASA is to keep these missions on schedule and within budget. The committee endorses NASA's policy of completing a mission's technological development before starting the mission. The committee similarly endorses such a policy as the NSF is applying it to the design and development of ALMA.

For possible ground-based initiatives in the decade 2010 to 2020, investment is required in very large, high-speed digital correlators; in infrared interferometry; and in specialized dark-matter detectors. Future space based initiatives require investment in spacecraft communication and x-ray interferometry, as well as technology for the next-generation observatories. Such technology will include energy-resolving array detectors for optical, ultraviolet, and x-ray wavelengths; far-infrared array detectors; refrigerators; large, lightweight optics; and gamma-ray detectors.

ASTRONOMY'S ROLE IN EDUCATION

Because of its broad public appeal, astronomy has a unique role to play in education and public outreach. The committee recommends that the following steps be taken to exploit the potential of astronomy for enhancing education and public understanding of science:

- Expand and improve the opportunities for astronomers to engage in outreach to the K-12 community.
- Establish more pilot partnerships between departments of astronomy and education at a few universities to develop exemplary science courses for preservice teachers.
- Improve communication, planning, and coordination among federal programs that fund educational initiatives in astronomy.
- Increase investment toward improving public understanding of the achievements of all NSF-funded science and facilities, especially in the area of astronomy.

1

Recommendations

INTRODUCTION

ASTRONOMY AND ASTROPHYSICS IN THE NEW MILLENNIUM

The new millennium marks a turning point in the history of efforts to understand our place in the universe. The heavens have been a source of fascination for humanity for thousands of years, but only in the last few centuries has it been possible to take the measure of the stars—and only in the last few decades, to take the measure of the entire universe. The scientific and technical revolution that has enabled this enormous progress is accelerating. We can anticipate major new discoveries in the early decades of the new millennium, but more importantly we can anticipate major advances in our *understanding* of the universe—its origin, its evolution, its ability to support life, its destiny. What is the nature of the matter and energy in the universe? What happened at the dawn of the modern universe, when the first stars and galaxies formed? How are black holes formed? How do stars and planets form, and how do planets evolve to create habitats suitable for life? How does the astronomical environment affect Earth? These questions are all part of the fabric of science, cutting across traditional disciplines and government agencies and connecting the universe from the smallest to the largest scales. Addressing them will require interactions of astronomy with many other disciplines, including physics, mathematics, computer science, and biology. The interaction with physics is particularly important since all objects in the universe—and indeed the universe itself—are governed by the same fundamental physical laws. Answering these questions will alter the perception of our place in the universe, just as the advent of the heliocentric theory did centuries ago. The search for the answers can also capture the imagination of the public and inspire interest in science, thereby helping to create a more scientifically literate citizenry.

ACCOMPLISHMENTS OF THE 1990s

The past decade saw an unprecedented number of important astronomical discoveries. Some highlights include:

- Discovery of planets orbiting other stars; about three dozen are now known.

- Determination of the interior structure of the Sun from observations of its seismic activity. These results confirmed theoretical models of solar structure throughout most of the Sun to within 0.1 percent and support the hypothesis that the observed deficit in the number of neutrinos from the Sun arises because they have a non-zero mass.
- Discovery of the Kuiper Belt, a large group of small, primitive bodies in the outer solar system predicted by theory. The Kuiper Belt is probably the source of most short-period comets and contains a unique fossil record of how the solar system formed.
- Observation of the impact of Comet Shoemaker-Levy 9 on Jupiter, providing a dramatic illustration of the potential effects of such impacts on Earth.
- Discovery of "brown dwarfs," cool stars too small to sustain nuclear reactions in their interiors.
- Discovery of the theoretically predicted phenomenon of gravitational microlensing, in which the brightness of background stars is amplified by the gravitational effects of intervening objects of stellar mass.
- Discovery that gamma-ray bursts originate in the very distant universe and that they produce afterglows at other wavelengths, as had been predicted theoretically.
- Discovery of massive black holes in the nuclei of galaxies, including our own Milky Way, thereby confirming theoretical predictions that such massive black holes must be common.
- Discovery of young galaxies at redshifts greater than 3, revealing the dramatic evolution of galaxies from the early universe to the present.
- Discovery of theoretically predicted tiny fluctuations in the background radiation left over from the Big Bang on scales from 100 million to 10 billion light-years, the seeds of subsequent structure formation.
- Measurement of the expansion rate of the universe to an accuracy approaching 10 percent and determination that there is not enough matter to stop the expansion of the universe.
- At the end of the decade of the 1990s, evidence suggesting both that the universe is "flat," as expected in inflationary cosmologies, and that its expansion is accelerating owing to the presence of "dark energy."

THE LEGACY OF THE PREVIOUS DECADAL SURVEY

Planning and investment over the past decade are now providing ripe opportunities for advancements and discoveries in the near future. All of the large and many of the moderate programs recommended in the report of the previous Astronomy and Astrophysics Survey Committee, *The Decade of Discovery in Astronomy and Astrophysics* (NRC, 1991; known as the Bahcall report and referred to in the current report as the 1991 survey), have been initiated. The top priority, the Space Infrared Telescope Facility (SIRTF), is scheduled for launch in 2002. SIRTF is the last of the Great Observatories, begun with the Hubble Space Telescope in the 1980s and successfully continued with the Compton Gamma Ray Observatory and the Chandra X-ray Observatory in the 1990s. SIRTF's remarkable sensitivity in the infrared will enable it to make fundamental advances in the study of brown dwarfs, in tracing the formation and evolution of galaxies, and in better understanding quasars. The first of the two ground-based Gemini optical-infrared telescopes was completed in 1999, and the second is scheduled to open in 2001. Each telescope mirror has a diameter of 8 m. These telescopes will enable study of some of the same sources as SIRTF, but at shorter wavelengths and with much finer angular resolution. The Millimeter Array, which has become part of the more powerful Atacama Large Millimeter Array (ALMA), is nearing the end of its design phase with construction poised to start. ALMA will trace the processes by which interstellar gas turns into stars, on small scales in our galaxy and on much larger scales in the distant universe.

Adaptive optics, the 1991 survey's top-priority moderate program, advanced substantially during the 1990s and received a major boost with the selection of the Center for Adaptive Optics as a National Science Foundation (NSF) Science and Technology Center. By eliminating the smearing effects of the atmosphere, adaptive optics will enable astronomers to see fainter objects and far greater detail with existing ground-based telescopes. The Astrometric Interferometry Mission has evolved into the far more capable Space Interferometry Mission (SIM). Measuring positions on the sky with unprecedented precision, SIM will enable the discovery of planets much more similar to Earth in mass and orbit than those detectable now, and it should permit astronomers to survey the Milky Way Galaxy 1,000 times more accurately than is possible now. Through the National Optical Astronomy Observatories, the NSF has contributed to two new 4-m-class telescopes, providing powerful new tools for astronomical investigations. Studies of the cosmic microwave background radiation, the top priority of the Task Group on Space Astronomy and Astrophysics (NRC, 1997), are being pursued through NASA's Microwave Anisotropy Probe (MAP), the European Space Agency's Planck Surveyor mission, and a vigorous ground-based and suborbital program. These studies will determine the large-scale properties of the universe and the fundamental cosmological parameters with remarkable precision. The dramatic increases in computer power foreseen a decade ago have been realized, enabling enormous advances in astrophysical simulation.

We must begin planning now to take advantage of the discoveries these new facilities will make possible. Continued progress requires new ideas for the next generation of facilities and innovative plans for operating them most effectively.

APPROACH AND SCOPE

IMPLEMENTATION OF THE CHARGE

To help carry out its charge (stated in full in the preface), the committee established nine panels with a total of 105 members.[1] These panels developed the cases for all the projects and almost all the policies that were then evaluated and prioritized by the committee according to the criteria given below. The chairs of the panels participated in almost all the deliberations of the committee, but as advocates for their disciplines they did not vote on priorities. The majority of the panels were organized on the basis of observational technique rather than scientific subdiscipline, so as to facilitate selection of the projects to be considered by the committee. To treat issues that did not fit neatly into the panel organization, cross-panel working groups were established in four areas: (1) astronomical surveys, (2) extrasolar planets, (3) laboratory astrophysics, and (4) NSF-funded national observatories. The comments of these working groups provided input to the panels and to the committee. The committee also benefited from the National Research Council report *Federal Funding of Astronomical Research* (NRC, 2000), which addressed demographic and funding issues raised in previous astronomy surveys.

In their deliberations, the panels and the committee received input from a broad cross section of the community. Sessions were held at two meetings of the American Astronomical Society (AAS), each attended by more than 300 people. The AAS also organized the Decadal Issues Discussion Forum on the World Wide Web, which attracted a number of thoughtful contributions. More than 100 individuals made presentations to the panels. Members of the committee and panel chairs organized more than 20 meetings in their local institutions to gather further input. Hundreds of e-mail messages were received by the panels and the committee. Members of the committee consulted with representatives of the funding agencies and of the Office of Management and Budget and with congressional staff. Eight distinguished members of the international astronomical community[2] attended most of the second meeting of the committee, adding valuable insights to the deliberations. International astronomers contributed to the work of the panels as well.

The members of the panels and of the committee all share a primary expertise in astronomical science, and the scientific merit of the proposed programs was the primary basis for determining priorities. However, the committee also considered technical readiness, cost-effectiveness, impact on education and public outreach, and the relation to other projects, both in the United States and abroad.

The committee made a careful attempt to define the boundary between projects it would consider and those it would not. First, judging that the Explorer and Discovery programs at NASA are suitably peer-reviewed, the committee did not make any recommendations on individual projects in these programs. Second, for evaluation of two NASA projects under consideration that will make both in situ measurements and remote observations—Solar Probe and Interstellar Probe—the committee decided to defer to the Committee on Solar and Space Physics of the Space Studies Board. Third, in the case of the interdisciplinary field of particle astrophysics, the committee evaluated projects that use particles as tools for remote observation, but not projects that focus on the physics of the particles themselves. Finally, based on informal discussions with the Committee on Gravitational Physics of the Board on Physics and Astronomy, the committee did not evaluate possible future upgrades to the Laser Interferometer Gravitational-wave Observatory (LIGO) project. This project is discussed in the NRC report *Gravitational Physics: Exploring the Structure of Space and Time* (NRC, 1999).

PURPOSE AND CONTENT OF THE TWO VOLUMES

This astronomy and astrophysics survey includes the report of the survey committee (this volume) plus a separate volume, *Astronomy and Astrophysics in the New Millennium: Panel Reports* (NRC, 2001), comprising the reports of seven of the nine panels. The reports of the Panel on Astronomy Education and Policy and the Panel on Benefits to the Nation from Astronomy and Astrophysics have been incorporated into this volume. This volume reflects the consensus of the Astronomy and Astrophysics Survey Committee; each chapter in the *Panel Reports* reflects the consensus of the corresponding panel. Every effort has been made to ensure that the recommendations in the two volumes are consistent; if any discrepancy between the report of the committee and a report of a panel remains, the report of the committee takes precedence. The purpose of the panel reports is to describe the new initiatives in more detail, to provide more extensive justification for the priorities, and to give additional priorities, applying typically to small projects, appropriate for individual subfields of astronomy and astrophysics.

The committee's primary recommendations for new initiatives are summarized in the remainder of the present chapter. Chapter 2 describes the scientific case underlying each recommendation. Programs already in operation or that have been recommended in previous astronomy surveys are summarized in Chapter 3, which also gives full explanations of the proposed new initiatives. Chapters 4 through 6, respectively, describe the benefits that astronomy and astrophysics provide to the nation, discuss the role of astronomy in education, and offer policy recommendations aimed at maintaining the health of the discipline and enhancing its contributions to science in general. A glossary of technical terms and a list of abbreviations and acronyms are included in an appendix to this volume.

OPTIMIZING THE RETURN ON THE NATION'S INVESTMENT IN ASTRONOMY AND ASTROPHYSICS

The United States has made and continues to make a significant investment in exploring the universe. In exchange, the nation deserves the maximum scientific return for this investment and widespread dissemination of the results. The astronomy and astrophysics enterprise depends on highly trained and motivated people, on technologically sophisticated facilities and missions, and on institutions properly equipped to manage them. How can this system be optimized to address the frontier scientific problems most effectively, both now and in the future?

BALANCING NEW INITIATIVES WITH THE ONGOING PROGRAM

The first step in optimizing the program is to achieve a proper balance between new initiatives and the ongoing program. Congress specifically asked the committee to address this issue, and its response and discussion are given in Chapter 6. Many of the policy recommendations discussed there are directly relevant to the question of balance.

- **The committee reaffirms the recommendations of the 1991 Astronomy and Astrophysics Survey Committee (NRC, 1991) by endorsing the completion of the Space Infrared Telescope Facility (SIRTF), the Millimeter Array (MMA; now part of the Atacama Large Millimeter Array, or ALMA), the Stratospheric Observatory for Infrared Astronomy (SOFIA), and the Astrometric Interferometry Mission (now called the Space Interferometry Mission, or SIM). Consistent with the recommendations of the Task Group on Space Astronomy and Astrophysics (NRC, 1997), the committee stresses the importance of studying the cosmic microwave background with the Microwave Anisotropy Probe (MAP) mission, the European Planck Surveyor mission, and ground-based and balloon programs.**

A particular attraction of SIM is its dual capability: It enables both the detection of planets through narrow-angle astrometry and the mapping of the structure of our galaxy and nearby galaxies through wide-angle astrometry. It is critical that an accuracy of a few microarcseconds for wide-angle measurements be achieved in order to address a wide variety of fundamental problems throughout the decade.

The committee endorses U.S. participation in the European Far Infrared Space Telescope (FIRST), which gives U.S. astronomers the chance to observe the entire far-infrared and submillimeter regions of the spectrum with no impediments from the residual atmosphere that can hinder observing from airborne platforms. The committee also endorses NASA's decision to continue to operate the Hubble Space Telescope at a reduced cost until the end of the decade to maintain the capability for critical space-based ultraviolet and optical observations.

- **To achieve the full scientific potential of a new facility, it is essential that, prior to construction, funds be identified for operation of the facility, for renewal of its instrumentation,**

and for grants for data analysis and the development of associated theory.

NASA follows the spirit of this recommendation in large part by including the costs of Mission Operations and Data Analysis (MO&DA) in budgetary planning for new missions. The wisdom of this approach is manifest in the wealth of discoveries and new interpretations published during the prime years of NASA missions. The committee recommends that funds for closely related theory be included in MO&DA as well; a specific proposal for "Theory Challenges" is outlined below.

The committee further recommends that the NSF include funds for (1) operations, (2) facility instruments and other capabilities that enable full exploitation of the new facility, and (3) grants for the ground-breaking research—both observational and theoretical—enabled by the new facility during its early, highly productive years. These recommendations are consistent with those of the 1991 survey. Based on the experience with several recently completed facilities, the committee has budgeted operations at 7 percent of the capital cost per year and instrumentation at 3 percent per year for the first 5 years of operation (see Chapter 6); the actual percentages should be based on the particular circumstances of the individual facility. To enable observers and theorists to explore and develop the full capabilities of new facilities, the committee recommends budgeting "facility grants" for research associated with major facilities at about 3 percent of the capital cost per year for the first 5 years. A cost-effective and competitive grants program for moderate facilities requires a somewhat higher percentage, and the committee recommends that facility grants for such facilities be budgeted at about 5 percent per year. No facility grants are recommended for small projects since the funds available would be too small. Funds for operations, instrumentation, and facility grants for a period of 5 years are included in the committee's cost estimates for most ground-based initiatives (see the section below, "Proposed Priorities for Ground- and Space-Based Initiatives").

- **Adequate funding for unrestricted grants that provide broad support for research, students, and postdoctoral associates is required to ensure the future vitality of the field; therefore new initiatives should not be undertaken at the expense of the unrestricted grants program.**

A strong grants program is critical both to realize the full science value from state-of-the-art facilities and to ensure the future health of the field. On the basis of the NASA experience, the committee believes that grants associated with new ground-based facilities, as proposed above, should address the data analysis and science needs for those facilities as well. Such grants should also support some postdoctoral associates and graduate students. However, past experience shows that it is often the individual investigator grants that are not tied to a specific facility or program—the unrestricted grants—that support the innovative new research that drives the future directions of astronomy. Previous decadal astronomy surveys have also emphasized the importance of unrestricted grants in advancing the field.

STRENGTHENING GROUND-BASED ASTRONOMY AND ASTROPHYSICS

The United States is fortunate to enjoy substantial state and private funding for ground-based optical and infrared facilities and, to a lesser extent, radio and solar facilities. To optimize the return on federal investment in ground-based astronomy, the committee recommends that:

- **U.S. ground-based optical and infrared facilities, radio facilities, and solar facilities should each be viewed by the National Science Foundation and the astronomical community as a single integrated system drawing on both federal and nonfederal funding sources.**

The NSF must draw on the strengths of the independent observatories to achieve its goals in ground-based astronomy. Independent observations are playing their appropriate roles effectively in both radio and solar astronomy, but less so in optical and infrared astronomy. Several of the committee's recommendations should help to build the partnership between federal and independent observatories in both radio astronomy (the Combined Array for Research in Millimeter-wave Astronomy and the Square Kilometer Array technology development) and optical and infrared astronomy (the Telescope System Instrumentation Program).

- **Effective national organizations are essential to coordinate, and to ensure the success and efficiency of, these systems.**

These national organizations should work with the universities and independent observatories in developing the next generation of telescopes. The National Radio Astronomy Observatory (NRAO) and the

National Astronomy and Ionosphere Center (NAIC) currently fulfill the committee's recommended role for radio astronomy, and the National Solar Observatory (NSO) does so for the solar physics community. The National Optical Astronomy Observatories (NOAO) as currently functioning and overseen does not fulfill this role for the community of ground-based optical and infrared observers. A plan for the transition of this organization should be developed, and a high-level external review, using appropriate criteria, should be initiated (see Chapter 6).

- **As part of the effort to develop effective systems of telescope facilities, universities and independent observatories should work with the national organizations to develop coherent strategic plans for each system (optical, solar, and radio) and to develop those facilities that are too large or expensive to fit within the resources of a single institution or consortium. Universities should assume the responsibility for purchasing, instrumenting, and operating small telescopes needed for their students and faculty.**

Optimizing the overall program also requires a mechanism to achieve the correct balance among existing facilities to address the important scientific problems as they evolve.

- **Cross-disciplinary competitive reviews should be held about every 5 years for all NSF astronomy facilities. In these reviews, it should be standard policy to set priorities and consider possible closure or privatization. New facilities should undergo their first competitive review between 5 and 10 years after they become operational.**

NASA holds a senior review of its MO&DA programs every 2 or 3 years; the greater frequency is appropriate in view of the shorter lifetime of space missions. The recommended 5-year reviews should provide input to NSF's Division of Astronomical Sciences (AST), enabling it to adjust its entire portfolio to meet changing scientific priorities and to satisfy its strategic plan.

The Department of Energy (DOE) supports a broad range of programs in particle and nuclear astrophysics and in cosmology. Such investigations probe the fundamental forces and the nature of matter in ways that directly complement accelerator-based experiments and basic theoretical research. DOE research in plasma physics also has many synergies with astrophysics. The scientific payoff of these programs would be even stronger with a clearly articulated strategic plan for DOE's programs that involve astrophysics.

- **Given the increasing involvement of the Department of Energy in projects that involve astrophysics, the committee recommends that DOE develop a strategic plan for astrophysics that would lend programmatic coherence and facilitate coordination and cooperation with other agencies on science of mutual interest.**

ENSURING THE DIVERSITY OF NASA MISSIONS

NASA's program during the past decade has been extraordinarily successful. Three of the four Great Observatories are now operational, and the fourth is nearing readiness for launch. The Explorer program is very successful and has elicited many highly innovative, cost-effective proposals for small missions from the community. The committee endorses the continuation of a vigorous Explorer program by NASA.

Opportunities for moderate-scale missions are less readily available, however. In the past, there were a number of extraordinarily successful moderate-sized missions, including the Infrared Astronomical Satellite (IRAS), the Cosmic Background Explorer (COBE), the International Ultraviolet Explorer (IUE), the Extreme Ultraviolet Explorer (EUVE), and the Rossi X-ray Timing Explorer (RXTE). Moderate missions such as these no longer fit within the cost cap of the Explorer program.

- **NASA should continue to encourage the development of a diverse range of mission sizes, including small, moderate, and major, to ensure the most effective returns from the U.S. space program.**

Consistent with this recommendation, several moderate space missions are recommended for this decade.

INTEGRATING THEORY CHALLENGES WITH NEW INITIATIVES

Astronomy advances by innovation in instrumentation, observation, and theory. This report recommends a number of projects with technologically advanced instrumentation that will enable observers to extend the frontiers of knowledge. In many instances, astrophysical theorists provide the ideas that guide the choice of instrumentation, the decisions about what to observe, and the interpretation of data. Adequate support of theory is therefore essential in optimizing the nation's investment in astronomy and astrophysics.

- **To encourage theorists to contribute to the planning of missions and facilities and to the interpretation and understanding of the results, one or more explicitly funded theory challenges should be integrated with most modcrate or major initiatives.**

A theory challenge is an initiative to focus the attention of the research community on a theoretical problem or area, broadly or narrowly defined, that is ripe for theoretical progress and relevant to a particular mission or facility. For ground-based projects, the theory challenges would be part of the associated facility grants; as pointed out above, small ground-based projects do not have facility grants and thus would not have theory challenges. By encouraging coordinated efforts to addresss broad theoretical problems relevant to new programs, this recommendation provides the opportunity for a qualitatively new approach to advancing researchers' conceptual understanding. Numerical simulation will play an important role in the theory challenges, both in advancing our understanding and in enabling detailed comparison with observation. The theory challenges should significantly enhance the effectiveness of theoretical research related to specific missions. The committee believes that broadly based theoretical research not tied to specific missions is also vital to astrophysics, and it should continue to be supported by the unrestricted grants program at the NSF and by the Astrophysics Theory Program at NASA. Indeed, one of the committee's recommended small initiatives is to augment the NASA Astrophysics Theory Program.

COORDINATING PROGRAMS AMONG FEDERAL AGENCIES

The section "Congressional Questions" in Chapter 6 of this report addresses several specific questions posed by the House of Representatives Science Committee (as specified in the committee's charge; see the preface). Among these is the question of an optimal strategy for the coordination of astronomical initiatives funded through NASA and NSF, and perhaps other federal agencies. The enormous scale of many astronomical problems requires a coordinated national approach. In many cases, investigations that span different wavelength bands and disciplinary boundaries are needed in order to achieve a fundamental understanding of the phenomena under study. Interagency coordination and cooperation are often essential for such a multidisciplinary approach.

Both ground- and space-based facilities can be used to address the scientific themes identified by the committee as ripe for progress throughout this decade. Such facilities are traditionally supported by NSF and NASA, respectively. In addition, some of the important outstanding problems lie in the domain of DOE, including ones linked to particle physics, nuclear physics, and cosmology. The committee recommends that each agency build on its own unique capabilities while recognizing those of related agencies, taking steps toward collaborations that it believes will prove fruitful. Each agency should have a strategic plan for astronomy and astrophysics in place and should also have scientific committees (such as DOE and NSF's Scientific Assessment Group for Experiments in Non-Accelerator Physics [SAGENAP] and NASA's Space Science Advisory Committee [SSAC]) available to evaluate proposed interagency collaborations. The Office of Science and Technology Policy could facilitate such interagency collaborations.

COLLABORATING WITH INTERNATIONAL PARTNERS

International collaboration enables projects that are too costly for the United States to carry out on its own, and it enhances the scientific return on projects by engaging the scientific and technical expertise of international partners. The Hubble Space Telescope (HST), the Gemini Telescope Project, and ALMA are all critically dependent on international collaboration, as is very long baseline interferometry, which by necessity is carried out using radio telescopes around the world. In many cases, international collaboration provides additional opportunities for U.S. astronomers to participate in major projects at the advancing edge of science, as in the case of the European Solar and Heliospheric Observatory, XMM-Newton, the Planck Surveyor and FIRST missions, and the Japanese Advanced Satellite for Cosmology and Astrophysics mission. Valuable opportunities for international collaboration exist for smaller missions as well. Many aspects of international collaboration are analyzed in the joint National Research Council-European Science Foundation report *U.S.-European Collaboration in Space Science* (NRC-ESF, 1998). The prerequisites for successful collaboration include (1) clear agreement on the financial commitments on each side; (2) close coordination in planning to minimize the risks associated with missed cost, schedule, or technical guidelines by any of the partners; (3) a significant, clearly defined contribution from each participant; and (4) a commitment to the free exchange of scientific data and results. Collaborations on major projects require the full support of the participating scientific communities, which can be ensured if the projects are among the very highest priorities of all of the participants.

To facilitate coordination in planning for future international collaborations, the committee invited a number of leading astronomers from other nations to its second meeting and engaged them in discussion of the projects under consideration for priority setting by the committee. To attempt to address some of the most challenging scientific questions facing them, astronomers will increasingly need "world facilities," which are so large that they require the participation of many nations to succeed. ALMA, a joint U.S., Canadian, European, and possibly Japanese project, will be the first true world facility in astronomy. The committee affirms the value of international collaboration for ground-based and space-based projects of all sizes. International collaboration plays a crucial role in a number of the programs recommended by this committee, including the Next Generation Space Telescope, the Expanded Very Large Array, the Gamma-ray Large Area Space Telescope, the Laser Interferometer Space Antenna, the Advanced Solar Telescope, and the Square Kilometer Array technology development. International participation in the 30-m-class Giant Segmented Mirror Telescope (GSMT) would offer many benefits. The significantly more ambitious 100-m "OWL" telescope under development by the European Southern Observatory represents an excellent opportunity for shared technology development and possible eventual U.S. collaboration. Terrestrial Planet Finder (TPF) offers a promising opportunity for collaboration with the European Space Agency (ESA), which is considering a similar mission called Darwin.

NEW INVESTMENTS IN ASTRONOMY AND ASTROPHYSICS

PROPOSED PRIORITIES FOR GROUND- AND SPACE-BASED INITIATIVES

Despite the enormous progress in astronomy and astrophysics in the past decade, many mysteries remain. How did the universe begin? Recent evidence indicates that the expansion of the universe is accelerating—what is the nature of the "dark energy" causing the acceleration? Most of the matter in the universe is invisible, and the nature of this dark matter remains a mystery. At present there is a vast gap in our knowledge of the evolution of the universe between the time at which the cosmic background radiation was produced, about 300,000 years after the Big Bang, and the time at which the most distant known galaxies emitted the light we see today, about a billion years later. This span of time includes the "dark ages," when the only radiation was the glow left over from the Big Bang, and the dawn of the modern universe, when the first stars and galaxies formed. Researchers believe that supermassive black holes formed at about the same time that galaxies did, but nothing is known about how this occurred. Much smaller black holes are forming even today by processes that are poorly understood. Star formation drives the evolution of galaxies and leads to planet formation, yet there are far more questions than answers about how this process works. The discovery of extrasolar planets in the past decade has presented many new mysteries, since all the planetary systems observed so far are completely different from our solar system. How did such planetary systems form and evolve? What is their relation to our solar system, and which is the norm? Finally, when we look at our own Sun, we find that its light varies slightly with time. These variations may have significant effects on Earth's climate, yet they are not now understood. To address these pressing scientific issues, and many others, the committee developed a comprehensive set of new initiatives in astronomy and astrophysics that will both vastly increase our knowledge of the universe and lead to many new discoveries. More important, these initiatives should enable us to achieve a greater understanding of the complex phenomena leading from the origin of the universe in the Big Bang to the existence of a life-bearing planet like Earth.

- **The Astronomy and Astrophysics Survey Committee recommends the approval and funding of the prioritized new initiatives listed in Table 1.1.**

Table 1.1 presents the priorities for initiatives for the decade 2000 to 2010. The priorities are listed separately for ground-based and space-based initiatives, both because the funding agencies are different (NSF, DOE, and DOD primarily for ground; NASA primarily for space) and because space facilities are intrinsically more expensive. Major and moderate projects are prioritized separately. The small initiatives consist

TABLE 1.1 Prioritized Initiatives and Estimated Federal Costs for the Decade 2000 to 2010

Ground-based[a]	Cost[b] ($M)	Space-based[c]	Cost[b] ($M)
Major Initiatives			
Giant Segmented Mirror Telescope (GSMT)[d]	350	Next Generation Space Telescope (NGST)[d]	1,000
Expanded Very Large Array (EVLA)[d]	140	Constellation-X Observatory (Con-X)	800
Large-aperture Synoptic Survey Telescope (LSST)	170	Terrestrial Planet Finder (TPF)[e]	200
		Single Aperture Far Infrared (SAFIR) Observatory[e]	100
Subtotal ground-based	660	Subtotal space-based	2,100
Moderate Initiatives			
Telescope System Instrumentation Program (TSIP)	50	Gamma-ray Large Area Space Telescope (GLAST)[d]	300
Advanced Solar Telescope (AST)[d]	60	Laser Interferometer Space Antenna (LISA)[d]	250
Square Kilometer Array (SKA) technology development	22	Solar Dynamics Observatory (SDO)	300
Combined Array for Research in Millimeter-wave Astronomy (CARMA)[d]	11	Energetic X-ray Imaging Survey Telescope (EXIST)	150
Very Energetic Radiation Imaging Telescope Array System (VERITAS)	35	Advanced Radio Interferometry between Space and Earth (ARISE)	350
Frequency Agile Solar Radio telescope (FASR)	26		
South Pole Submillimeter-wave Telescope (SPST)	50		
Subtotal ground-based	254	Subtotal space-based	1,350
Small Initiatives			
National Virtual Observatory (NVO)	15	National Virtual Observatory (NVO)	45
Laboratory Astrophysics Program	5	Advanced Cosmic-ray Composition Experiment for the Space Station (ACCESS)	100
Low Frequency Array (LOFAR)	8		
National Astrophysical Theory Postdoctoral Program	6	Augmentation of the Astrophysics Theory Program	30
Synoptic Optical Long-term Investigation of the Sun (SOLIS) expansion	8	Laboratory Astrophysics Program	40
		National Astrophysical Theory Postdoctoral Program	14
		Ultralong-Duration Balloon Program	35
Subtotal ground-based	42	Subtotal space-based	264
Total ground-based	956	Total space-based	3,714
DECADE TOTAL			4,670

[a]Cost estimates for ground-based capital projects include technology development plus funds for operations, new instrumentation, and facility grants for 5 years.

[b]Best available estimated costs to U.S. government agencies in millions of FY2000 dollars and rounded. Full costs are given for all initiatives except TPF and the SAFIR Observatory.

[c]Cost estimates for space-based projects exclude technology development.

[d]Cost estimate for this initiative assumes significant additional funding to be provided by international or private partner; see *Astronomy and Astrophysics in the New Millennium: Panel Reports* (NRC, 2001) for details.

[e]These missions could start at the turn of the decade. The committee attributes $200 million of the $1,700 million total estimated cost of TPF to the current decade and $100 million of the $600 million total estimated cost of the SAFIR Observatory to the current decade.

of programs such as the National Virtual Observatory (NVO) and the Laboratory Astrophysics Program, plus illustrative small facilities and missions. The NVO is the top priority among these initiatives. The remaining ones are not prioritized, since small projects often have a gestation time of less than a decade. In particular, the committee has not recommended any projects for NASA's extremely successful Explorer program, since missions are selected through competitive peer review.

The size categories for new initiatives are based on the capital cost for ground-based projects and on the total cost, excluding technology development, for space-based projects. Only costs to be borne by the federal government are included. The committee's cost estimates for these initiatives are based on discussions with agency personnel and on presentations to the panels; they are given in FY2000 dollars. For ground-based projects, small projects have capital costs of up to $5 million; moderate, from $5 million to $50 million; and major, above $50 million. In contrast to the practice in previous astronomy and astrophysics decadal surveys, the tabulated costs for ground-based capital projects include operations and new instrumentation for a period of 5 years at rates of 7 percent and 3 percent of the capital cost per year. In addition, grants for data analysis and associated theory are included at a rate of 3 percent of the capital cost per year for major projects, 5 percent for moderate projects, and 0 percent for small projects. The total costs for ground-based initiatives are thus typically 1.65, 1.75, and 1.50 times the capital costs for major, moderate, and small initiatives, respectively. Exceptions are SKA technology development, which includes only funds for a theory challenge, budgeted at $200,000 per year for the decade; the Telescope System Instrumentation Program, which as an instrumentation program does not require operations or instrumentation funds and is too fragmented to have a grants program; and NVO, the National Astrophysical Theory Postdoctoral Program, and the Laboratory Astrophysics Program, which are not capital projects and therefore have no added costs. The Large-area Synoptic Survey Telescope is expected to have significant expenses for data analysis, and so the estimate of the total operations cost by the Panel on Optical and Infrared Astronomy from the Ground (see *Panel Reports;* NRC, 2001) was used for this project.

The costs of space-based initiatives given in Table 1.1 do not include technology development. NASA has adopted a policy of deferring construction of new missions until all major technological problems have been solved, a policy the committee endorses. These costs amount to typically about 30 percent of the construction costs of a mission. In some cases, entire missions will serve as precursors for other missions, for example, SIM for TPF as currently envisioned. Explorer and Discovery missions are regarded as small initiatives. Since they are peer-reviewed, the committee did not prioritize them. Moderate space-based missions are those with construction, launch, and operations costs between the $140 million cap on Explorer missions and $500 million; major missions have estimated costs above $500 million.

Many of the projects listed in Table 1.1 have been studied extensively and should have reasonably accurate cost estimates (see the discussions of individual projects in the *Panel Reports;* NRC, 2001). For others, such as TPF and GSMT, more accurate cost estimates must await the outcome of future technology developments. To compare the total cost of the recommended new initiatives with that of the initiatives recommended in the 1991 survey, it is first necessary to correct for inflation. In FY2000 dollars, the 1991 survey recommended about $640 million for ground-based initiatives and $3.3 billion for space-based initiatives, for a total of $3.9 billion. The total cost of the current committee's recommendations for ground-based astronomy, $956 million, is higher than that of the 1991 survey's recommendations because it includes 5 years' worth of operating costs, new instruments, and grants—expenses that were not included in the 1991 survey's estimates. If these costs are removed, the committee's recommendation for ground-based astronomy is about $600 million, slightly less than that of the 1991 survey.

The estimated decade cost of the committee's recommended initiatives for space-based astronomy is about $3.7 billion. Two of the projects, TPF and the SAFIR Observatory, could start around the end of this decade. TPF is included in NASA's strategic plan for this decade and is under active development at this time; however, the technical challenges facing this mission could delay its start until after 2010. The SAFIR Observatory could also start in this decade, but delays in its precursor mission, NGST, could push SAFIR's start into the following decade. The committee believes that it is essential that all missions be subject to timely review independent of their phasing with respect to decadal surveys, and for this reason it has prioritized both missions. The committee notes that NASA was so successful in addressing the space-based initiatives recommended by the 1991 survey that the NRC was asked to convene the Task Group on Space Astronomy and Astrophysics in the mid-1990s to prioritize science for further NASA missions (NRC, 1997). The committee assumed that about 15 percent of the total cost of TPF and SAFIR will fall in this decade.

The committee's priorities for major and moderate programs, independent of whether they are ground- or space-based, and independent of funding agency, are presented in Table 1.2.

EXPLANATION OF NEW INITIATIVES

Brief descriptions of the recommended new initiatives are given here, in priority order for each size category. More detailed information can be found in Chapter 3 and in the *Panel Reports* (NRC, 2001)

MAJOR INITIATIVES

Next Generation Space Telescope. NGST is the top priority for this decade because it will reveal the first epoch of star formation and trace the evolution of galaxies from their birth to the present. It will also provide a unique window onto the birth of stars and planets in our own galaxy. Having NGST's sensitivity extend to 27 μm would substantially improve its ability to study Kuiper Belt objects (KBOs) in our solar system, the formation of stars and planets in our galaxy, and the dust emission from galaxies out to redshifts of 3. As an 8-m-class passively cooled space telescope, it will be more than 100 times as sensitive as HST or SIRTF in the infrared and will improve image sharpness by an order of magnitude. Its potential for new discoveries will easily rival that of HST when it was launched. The European Space Agency and the Canadian Space Agency plan to make substantial contributions to the instrumentation of NGST.

Giant Segmented Mirror Telescope. The second priority overall, and the top priority for ground-based astronomy, is to develop the technology for and begin construction of a giant (30-m-class) segmented-mirror telescope equipped with adaptive optics. GSMT, with its higher spatial resolving power and its greater capability for high-resolution spectroscopy, will be a powerful complement to NGST in tracing the evolution of galaxies and in studying the formation of stars and planets. It will exceed NGST in sensitivity below a wavelength of 2 μm, but atmospheric and thermal effects will compromise its sensitivity at longer wavelengths. GSMT will be a uniquely powerful instrument for studies of the evolution of the intergalactic medium and the history of star formation in the Milky Way Galaxy and its nearest neighbors. The ability to add new instruments means that GSMT will be able to respond to new scientific opportunities and take advantage of novel technological developments in

TABLE 1.2 Prioritized Initiatives (Combined Ground and Space) and Estimated Federal Costs for the Decade 2000 to 2010[a,b]

Initiative	Cost[c] ($M)
Major Initiatives	
Next Generation Space Telescope (NGST)[d]	1,000
Giant Segmented Mirror Telescope (GSMT)[d]	350
Constellation-X Observatory (Con-X)	800
Expanded Very Large Array (EVLA)[d]	140
Large-aperture Synoptic Survey Telescope (LSST)	170
Terrestrial Planet Finder (TPF)[e]	200
Single Aperture Far Infrared (SAFIR) Observatory[e]	100
Subtotal for major initiatives	2,760
Moderate Initiatives	
Telescope System Instrumentation Program (TSIP)	50
Gamma-ray Large Area Space Telescope (GLAST)[d]	300
Laser Interferometer Space Antenna (LISA)[d]	250
Advanced Solar Telescope (AST)[d]	60
Square Kilometer Array (SKA) technology development	22
Solar Dynamics Observatory (SDO)	300
Combined Array for Research in Millimeter-wave Astronomy (CARMA)[d]	11
Energetic X-ray Imaging Survey Telescope (EXIST)	150
Very Energetic Radiation Imaging Telescope Array System (VERITAS)	35
Advanced Radio Interferometry between Space and Earth (ARISE)	350
Frequency Agile Solar Radio telescope (FASR)	26
South Pole Submillimeter-wave Telescope (SPST)	50
Subtotal for moderate initiatives	1,604
Small Initiatives	
National Virtual Observatory (NVO)	60
Other small initiatives	246
Subtotal for small initiatives	306
DECADE TOTAL	4,670

[a]Cost estimates for ground-based capital projects include technology development plus funds for operations, new instrumentation, and facility grants for 5 years.

[b]Cost estimates for space-based projects exclude technology development.

[c]Best available estimated costs to U.S. government agencies in millions of FY2000 dollars and rounded. Full costs are given for all initiatives except TPF and the SAFIR Observatory.

[d]Cost estimate for this initiative assumes significant additional funding to be provided by international or private partner; see *Panel Reports* (NRC, 2001) for details.

[e]These missions could start at the turn of the decade. The committee attributes $200 million of the $1,700 million total estimated cost of TPF to the current decade and $100 million of the $600 million total estimated cost of the SAFIR Observatory to the current decade.

instrumentation. The committee recommends an immediate start for the technology development needed to reduce the cost of construction and to develop the adaptive optics. The cost of technology development and construction is estimated to be about $400 million. It is assumed that half of these costs and half of the operations costs will be borne by private and/or international partners; the cost estimates in Tables 1.1 and 1.2 are therefore based on a federal capital cost of $200 million. Open access to GSMT by the U.S. astronomical community should be directly proportional to the investment by the NSF.

Constellation-X Observatory. The premier instrument to probe the formation and evolution of black holes—both stellar black holes in our galaxy and supermassive black holes in the nuclei of other galaxies—Constellation-X will also measure the physical conditions in the first clusters of galaxies, study quasars at high redshift, contribute to nuclear physics by measuring the radii of neutron stars, and trace the formation of the chemical elements. To achieve the sensitivity needed to meet these goals, Constellation-X will consist of four x-ray telescopes in separate spacecraft. Each telescope will have high spectral resolution over a broad energy range, ~ 0.25 to 40 keV. Constellation-X has been under active study for more than 5 years, and the technology issues are well in hand for a start in the middle of the decade.

Expanded Very Large Array. The VLA is currently the premier centimeter-wavelength radio telescope in the world, despite being based on the technology of 20 to 30 years ago. Replacing key components with modern technology will provide an order-of-magnitude increase in sensitivity with unprecedented image quality and a 1,000-fold increase in spectroscopic capability, all at a fraction of the cost of constructing a new facility. The addition of eight new antennas will provide an order-of-magnitude increase in angular resolution, making it comparable to that of ALMA and NGST. The EVLA will be a powerful instrument for studying the formation of protoplanetary disks and stars, as well as the formation and evolution of the first galaxies.

Large-aperture Synoptic Survey Telescope. By surveying the visible sky every week to a much fainter level than can be achieved with existing optical surveys, LSST will open a new frontier in addressing time-variable phenomena in astronomy. This 6.5-m-class optical telescope will detect 90 percent of the near-Earth objects larger than 300 m in diameter within a decade, and will enable assessment of the potential hazard each poses to Earth. It will take a census of some 10,000 of the most primitive bodies in the solar system, located in the Kuiper Belt. It will also contribute to the study of the structure of the universe by observing thousands of supernovae, both nearby and at large redshift, and by measuring the distribution of dark matter through gravitational lensing. LSST will produce a terabyte of data per night, all of which will be accessible to scientists and the public alike through the National Virtual Observatory.

Terrestrial Planet Finder. The main goal of TPF is nothing less than to search for evidence of life on terrestrial planets around nearby stars. The present concept calls for a space-based infrared interferometer of enormous sensitivity, capable of nulling out the light from the host star. TPF's angular resolution will also enable it to peer into the innermost regions of protoplanetary disks, galactic nuclei, starburst galaxies, and galaxies at high redshift. By a large margin, TPF is the most costly and the most technically challenging mission discussed in this report. Both SIM and NGST involve key technologies that must be successfully demonstrated if TPF as currently envisioned is to go forward. The committee's recommendation of this mission is predicated on the assumptions that TPF will revolutionize major areas of both planetary and nonplanetary science, and that, prior to the start of TPF, ground- and space-based searches will confirm the expectation that terrestrial planets are common around solar-type stars. NASA should pursue a vigorous program of technology development to enable the construction of TPF to begin in this decade.

Single Aperture Far Infrared Observatory. The SAFIR Observatory will take advantage of the technology developed for NGST to study the relatively unexplored region of the spectrum between 30 and 300 μm. It will investigate the earliest stage of star formation and galaxy formation by revealing regions too shrouded by dust to be studied with NGST and too warm to be studied effectively with ALMA. An 8-m-class space-based telescope that is diffraction-limited at 30 μm, it will be more than 100 times as sensitive as SIRTF or the European FIRST mission. It will have the capability of becoming part of an interferometer at a later time.

MODERATE INITIATIVES

Telescope System Instrumentation Program. Universities and independent observatories operate the majority of the large optical and infrared telescopes in the United States. The top priority for moderate projects for the new decade is TSIP, an NSF program that will maximize the scientific effectiveness of these telescopes by instrumenting them and by making them accessible to the entire astronomical community. It will have a multiplier effect by encouraging the continuation of substantial nonfederal investments, while at the same time helping to bring the national and private observatories together as a coherent research system. Under this initiative, the NSF would fund the construction of peer-reviewed new instrumentation at private observatories in exchange for telescope time or other equally valuable benefits for the community at large. The value of the observing time (determined from the cost of operations and the amortized investment) should be 50 percent of the granted funds.

Gamma-ray Large Area Space Telescope. GLAST, the second priority for moderate projects, is a powerful gamma-ray telescope with a sensitivity 30 times greater than that of its predecessor, the Energetic Gamma Ray Experiment (EGRET) instrument on the Compton Gamma Ray Observatory. GLAST will study powerful jets from the supermassive black holes in the centers of distant galaxies, the acceleration mechanisms of cosmic rays, and the origin of tremendous bursts of gamma-ray radiation from the distant universe. GLAST has the potential for breakthrough discoveries, such as by observing gamma rays from dark matter annihilation. It will detect gamma rays in the photon energy range between 10 MeV and 300 GeV with unprecedented positional accuracy. The committee applauds the crucial technical contributions of DOE to this important NASA mission.

Laser Interferometer Space Antenna. LISA is unique among the recommended new initiatives in that it is designed to detect the gravitational radiation predicted by Einstein's theory of general relativity. The direct measurement of gravitational radiation from astrophysical sources will open a new window onto the universe and enable investigations of the physics of strong gravitational fields. LISA consists of three spacecraft spaced 5 million km apart in an equilateral triangle, with lasers accurately monitoring their separation. It will be sensitive to gravitational-wave frequencies between 10^{-1} to 10^{-4} Hz, frequencies too low to be detected by the ground-based LIGO. For the first time, it will be possible to observe evidence of the coalescence of supermassive black holes as distant galaxies merge, and the gravitational radiation from white dwarf binaries in our own galaxy. It is assumed that LISA will be a joint mission between NASA and ESA, with costs shared approximately equally.

Advanced Solar Telescope. AST will observe solar plasma processes and magnetic fields with unprecedented resolution in space and time. It will provide critical information needed to solve the mysteries associated with the generation, structure, and dynamics of the surface magnetic fields, which govern the solar wind, solar flares, and short-term solar variability. AST is a ground-based, 4-m-class, adaptive-optics-equipped facility that will operate in the wavelength band from 0.3 to 35 μm. It is proposed as a joint project with international partners, in which the United States would provide about half the costs.

Square Kilometer Array Technology Development. The SKA is an international ground-based centimeter-wave radio telescope array with 10^6 square meters of collecting area that will enable study of the first structures and the first luminous objects to form during the dawn of the modern universe, and will provide unprecedented images of protostellar disks and the neutral jets launched by young stars. SKA's sensitivity will be a factor of 100 greater than that of existing centimeter-wave facilities. The increase in sensitivity has great discovery potential, and SKA will revolutionize the study of objects and phenomena that are currently undetectable at centimeter wavelengths. The U.S. SKA development program will, in collaboration with the international radio astronomy community, aggressively pursue technology and technique development in this decade that will enable the construction of the SKA in the following decade.

Solar Dynamics Observatory. SDO will probe the outer layers of the Sun to determine the connections between the interior dynamics and the activity of the solar corona, the origin of sunspots and solar active regions, and the origin of coronal mass ejections and solar flares. SDO is a space-based mission that will carry a number of instruments and small telescopes to monitor the Sun continuously at ultraviolet and optical wavelengths (0.02 to 1 μm).

Combined Array for Research in Millimeter-wave Astronomy. CARMA is the planned combination of the Berkeley-Illinois-Maryland Association (BIMA) and the Owens Valley Radio Observatory (OVRO) millimeter-wave arrays at a superior site along with the addition of new, smaller antennas. The resulting hybrid array will have unique imaging qualities, sensitive to both small and large scales. It will study star formation at all epochs, and it will measure the small distortions in the cosmic (Big Bang) microwave background caused by the hot gas in distant clusters of galaxies along the line of sight. Its Northern Hemisphere location will provide sky coverage complementary to ALMA. Of the construction costs, 60 percent will come from nonfederal sources.

Energetic X-ray Imaging Survey Telescope. EXIST will survey the entire sky every 90 minutes to search for weak and often time-variable astronomical sources of 5- to 600-keV x-ray photons. Such x rays emanate from many sources, including supermassive black holes in the centers of galaxies, stellar mass black holes, neutron stars, and embedded supernovae in our galaxy, and the mysterious distant sources of gamma-ray bursts of radiation. Attached to the International Space Station, EXIST will survey sources 1,000 times weaker than the sources in the previous hard x-ray survey by the High Energy Astronomical Observatory (HEAO-1). EXIST's repeated surveys of the entire sky in the hard x-ray region will complement those by LSST at optical wavelengths.

Very Energetic Radiation Imaging Telescope Array System. VERITAS will perform the first sensitive sky survey for astronomical sources of extremely energetic photons—those with energies from 100 to 10,000 GeV. VERITAS will complement GLAST and EXIST in studying the cosmic sources of relativistic particles such as supermassive black holes, gamma-ray burst sources, pulsars, and supernova remnants. Making use of the established technology of the 10-m reflector at the Whipple Observatory, VERITAS consists of an array of seven 10-m-diameter reflectors that will achieve more than an order-of-magnitude improvement in sensitivity and have a far greater ability than existing instruments to locate sources.

Advanced Radio Interferometry between Space and Earth. ARISE is a sensitive, Earth-orbiting radio antenna with a diameter of about 25 m that will improve the angular resolution of the ground-based Very Long Baseline Array by a factor 6. It will probe the regions near supermassive black holes, which are thought to produce relativistic jets, and will enable the study of maser sources, both in the Milky Way and in the nuclei of other galaxies.

Frequency Agile Solar Radio Telescope. Radio waves from the Sun bring information about the heating of the corona, the nature and evolution of coronal magnetic fields, the structure of the solar atmosphere, and the origin of the solar wind. FASR will analyze this radiation over a frequency range from 0.3 to 30 GHz, and it will improve on existing facilities by operating at hundreds of frequencies and providing a factor-of-10 better spatial resolution.

South Pole Submillimeter-wave Telescope. The SPST will take advantage of the superb atmospheric transmission conditions between wavelengths of 200 μm and 1 mm at the South Pole to survey the dusty universe, study small variations in the background radiation emanating from the early pregalactic universe, and identify primordial galaxies.

Other Projects

Two areas in which the committee considered it premature to set priorities are cosmic microwave background experiments and particle astrophysics. In both disciplines, projects are often experiments rather than observatories. As a result, it is difficult to plan far in advance, since the best strategy may depend critically on the outcome of ongoing experiments.

Observations of the cosmic microwave background can determine the large-scale properties of the universe and reveal the tiny fluctuations that were the seeds of all the structure we see today. These observations are of fundamental importance to both astronomy and physics. Together with ground-based and balloon experiments, NASA's MAP mission, to be launched in spring 2001, will revolutionize knowledge of the microwave background, and the committee believes that no decision on the next major or moderate microwave background project should be made until the results from that mission are available. ESA's Planck mission later in the decade will also provide important information, but it will be possible to decide on the next step before its results are available. Together, MAP and Planck will test the most promising ideas about the very early universe as well as determine cosmological parameters to high precision. The next frontier is to measure the polarization of the cosmic microwave background, which has the potential of probing even earlier times, close to the Big Bang itself.

Particle astrophysicists view the universe through different eyes than do most astronomers: Instead of photons, they observe energetic particles, and they search for exotic new particles that could account for the dark matter. The committee decided that it was too early to prioritize the Northern Hemisphere Auger project or Telescope Array for ultrahigh-energy cosmic rays and the Ice Cube experiment for high-energy neutrinos. In each case, ongoing experiments—the Southern Hemisphere Pierre Auger Observatory project and the Antarctic Muon and Neutrino Detector Array (AMANDA) experiment at the South Pole—will provide critical information that will enable an informed decision to be made soon. The decision on how to pursue future searches for dark matter will be guided by the outcome of experiments now just starting.

Small Initiatives

National Virtual Observatory. The NVO is the committee's top-priority small initiative. NVO involves the integration of all major astronomical data archives into a digital database stored on a network of computers, the provision of advanced data exploration services for the astronomical community, and the development of data standards and tools for data mining. It will create a powerful resource for public education and outreach by making near-real-time observations accessible over the Internet. NVO will enable professional astronomers, educators, and the public to take full advantage of the wealth of data from existing and planned surveys such as LSST. NVO is made possible by huge advances in the past decade in computer speed, widespread access to high-speed networks, a dramatic decrease in the cost of computing and data storage capabilities, and an ongoing revolution in techniques to extract science from a large data set. The committee recommends coordinated support from both NASA and the NSF, since NVO will serve both the space- and ground-based science communities.

Additional Small Intiatives. The remaining recommendations for small initiatives are listed alphabetically and are not prioritized, but they represent a number of exciting opportunities for making a relatively small investment to potentially achieve a major gain in capability and scientific return. Several of these opportunities span both space- and ground-based astronomy, like NVO. To facilitate the interpretation of the tremendous harvest of data throughout the decade, the committee recommends (1) augmentation of NASA's Astrophysics Theory Program and (2) initiation of a NASA- and NSF-funded National Astrophysical

Theory Postdoctoral Program that would award 10 portable 3-year postdoctoral positions each year. In view of the tremendous increase in the volume and complexity of the spectroscopic data to be gathered across the electromagnetic spectrum by new facilities over the course of the decade, the committee recommends the establishment of a Laboratory Astrophysics Program funded by both NASA and the NSF. The Ultralong-Duration Balloon (ULDB) program offers the prospect of carrying payloads that weigh several tons to roughly 40-km altitudes for flights that last several months. For certain types of experiments, this program offers an alternative to small space experiments at a fraction of their cost. The committee recommends an augmentation of the ULDB program for the development of a steering capability that would increase both the duration and the probability of success of the balloon flights. The Advanced Cosmic-ray Composition Experiment for the Space Station (ACCESS) will measure the spectrum and composition of cosmic rays with energies up to 1,000 TeV. This experiment will provide unique data for studying the origin of cosmic rays and the mechanism by which they are accelerated. The Low Frequency Array (LOFAR), a joint project now under way between the Netherlands Foundation for Research in Astronomy and the U.S. Naval Research Laboratory, provides a 100- to 1,000-fold improvement in sensitivity and resolution over existing radio telescopes in the wavelength range from 2 to 20 m. It has the potential to discover the hydrogen clouds out of which the first galaxies formed. The committee recommends that the NSF contribute to LOFAR so that it will be available to the entire astronomical community. Expansion of the Synoptic Optical Long-term Investigation of the Sun (SOLIS) from one station to a three-station network would permit continuous, 24-hour monitoring of magnetic fields on the solar surface. An expanded SOLIS promises crucial data for understanding the magnetic origin of solar variability and a far greater capability to forecast the space weather that so adversely affects space satellites.

TECHNOLOGY

Technological innovation has often enabled astronomical discovery. Most of the major discoveries listed at the beginning of this chapter were possible only because of the remarkable advances in technology in the past two decades. Continued investment in technology in this decade is required for many of the initiatives recommended in this report: For example, GSMT and AST require advances in adaptive optics, and TPF requires the development of space interferometry. For the space-based initiatives, technology investment as specified in the existing NASA technology road map is an assumed prerequisite for the cost estimates in Table 1.1. It is essential to maintain funding for the planned technology development if NASA is to keep these missions on schedule and within budget. Targeted technology programs involving a joint effort between engineers and scientists will be essential to success in these projects. As noted above, the committee endorses NASA's policy of completing the technological development of a mission prior to starting it. The NSF is applying a similar approach the design and development of ALMA, a policy the committee endorses.

Longer-range investments in technology in this decade are needed to enable the major projects in the next decade—and to make them more cost-effective. Space-based projects that could be started in the decade 2010 to 2020 include a large 8-m-class ultraviolet telescope, a far-infrared interferometer, an x-ray interferometer, and an x-ray telescope with an effective area of 100 square meters. The best-defined ground-based project for the decade 2010 to 2020 is the Square Kilometer Array, for which the committee has recommended technology development. Other possible ground-based projects for that decade include the next step beyond GSMT in optical telescopes and large interferometers that operate at infrared wavelengths.

To make these projects feasible in the decade 2010 to 2020, the committee recommends investing in the technologies listed in Table 1.3; priorities for these technologies have not been established. For ground-based astronomy, the committee recommends development of very large, high-speed digital correlators for radio astronomy; infrared interferometry; and specialized dark-matter detectors. Space-based astronomy requires investments in spacecraft communication to enable high rates of data transmission from distant telescopes to ground-based stations. The estimated cost for the development of a suitable radio transmitter is listed in Table 1.3. In addition, the committee suggests that NASA consider establishing an optical communications link at "L2," the proposed site for NGST and other future NASA missions. The committee recommends investing in the development of x-ray interferometry, which has the potential of actually imaging the event horizon of a black hole, and in technology for the next generation of space observatories: energy-resolving array detectors for optical, ultraviolet, and x-ray wavelengths; far-infrared array detectors; refrigerators to maintain the cryogenic temperatures needed by these detectors; large, lightweight optics (some-

TABLE 1.3 Technology Development for Future Initiatives and Estimated Costs

Technology Initiative	Decade Cost ($M)
Ground-based	
Megacorrelators	10
Infrared interferometry	40
Dark matter detectors	12
Subtotal for ground-based technology	62
Space-based	
Spacecraft communication	70
X-ray interferometry	60
Technology for next-generation observatories:	
Energy-resolving array detectors	40
Far-infrared array detectors	10
Refrigerators for space experiments	50
Large, lightweight optics	80
MeV detector technology	10
Subtotal for space-based technology	320
TOTAL	382

times referred to as "gossamer optics") for infrared, optical, ultraviolet, and x-ray wavelengths; and sensitive gamma-ray (MeV) detectors. These proposed technology developments are discussed in greater detail in the *Panel Reports* (NRC, 2001).

ASTRONOMY'S ROLE IN EDUCATION

Astronomers have a vital role to play in contributing to the development of science education in the United States. Among scientists, astronomers make a disproportionately large contribution to the improvement of public science literacy relative to the comparatively small size of the astronomical community because of the broad appeal of astronomical concepts and ideas. Astronomy resonates with some of the most basic questions of humanity: When did the universe begin? How has it evolved? What will be its ultimate fate? Is there life elsewhere?

How will the universe around us affect the development and continued existence of the human species?

Astronomy provides a gateway for increased public understanding of humanity's place in the universe and of the nature of science. Each year 28 million visits are made to planetariums in the United States. More than 200,000 college students take an astronomy course each year. Thus, roughly 10 percent of all U.S. college students will take an astronomy course before they graduate, and for many of these astronomy will be the only science course they ever take. Astronomy is heavily covered by the media, has attracted millions of people on the Web, and enjoys the support of hundreds of thousands of amateur astronomers.

The committee recognizes the tremendous importance of improving the scientific literacy of the nation. Recommendations to enhance astronomy's role in education are discussed in Chapter 5. The key recommendations are as follows:

- **The engagement of astronomers in outreach to the K-12 community should be expanded and improved by ensuring (1) appropriate incentives for their involvement; (2) training and coordination for effective and high-leverage impact; (3) careful scrutiny of major initiatives and widespread dissemination of information regarding their successes and failures; and (4) recognition of the value of this work by the scientific community.**

- **More universities with both astronomy and education departments should establish pilot partnerships to bring scientists, educators, and experienced teachers together to design exemplary astronomy-based science courses for preservice teachers, with the goal of contributing to the achievement of long-term systemic reform in K-12 science education.**

- **Federal agencies charged with increasing the contribution of professional scientists to educational initiatives should work with astronomers and educators to develop a common set of goals, pathways to achieve them, and mutually accepted standards for measuring success.**

- **NSF should invest additional resources in improving public recognition of the achievements of all NSF-funded facilities and projects. Astronomy, with its wide public appeal, could provide a starting point.**

Achieving better recognition for NSF-sponsored successes would require that a stronger interface be established between the media and the NSF through the efforts of dedicated press officers. The NSF also should develop a stronger presence on the Internet. NSF centers and facilities should develop informative Web pages, maintain state-of-the-art visitors centers, and expand outreach into the local communities.

NOTES

1. Panel members are listed in the front matter.

2. Catherine Cesarsky, DSM Orme des Merisiers, now Director General, European Southern Observatory; Edward van den Heuvel, University of Amsterdam; Don Morton, Herzberg Institute of Astrophysics, National Research Council of Canada; Luis Rodriguez, Instituto de Astronomiá, National Autonomous University of Mexico; Yasuo Tanaka, Institute of Space and Astronautical Sciences, Japan, and Max-Planck-Institut für extraterrestrische Physik; Reinhard Genzel, Max-Planck-Institut für extraterrestrische Physik; Sami Solanki, ETH Zürich Institute of Astronomy, Switzerland; and Roger Davies, Durham University, United Kingdom.

2

The Science Behind the Recommendations

A VISION FOR ASTRONOMY AND ASTROPHYSICS IN THE NEW CENTURY

In the year 1000 AD there were astronomers in only a few places on Earth: in Asia, particularly China, in the Middle East, and in Mesoamerica. These astronomers were aware of only six of the nine planets that orbit the Sun. Although they studied the stars, they did not know that the stars were like the Sun, nor did they have any concept of their distances from Earth. By the year 2000 AD, humanity's horizons had expanded to include the entire universe. We now know that our Sun is but one of 100 billion stars in the Milky Way Galaxy, which is but one of about 100 billion galaxies in the visible universe. More remarkably, our telescopes have been able to peer billions of years into the past to see the universe when it was young—in one case, when it was only a few hundred thousand years old. All these observations can be interpreted in terms of the inflationary Big Bang theory, which describes how the universe has evolved since the first 10^{-36} seconds of cosmic time.

It is impossible to predict where astronomy will be in the year 3000 AD. But it is clear that for the foreseeable future, the defining questions for astronomy and astrophysics will be these:

- How did the universe begin, how did it evolve from the soup of elementary particles into the structures seen today, and what is its destiny?
- How do galaxies form and evolve?
- How do stars form and evolve?
- How do planets form and evolve?
- Is there life elsewhere in the universe?

Researchers now have at least the beginnings of observational data that are relevant to all of these questions. However, a relatively complete answer exists for only one of them—how stars evolve. The development and observational validation of the theory of stellar evolution was one of the great triumphs of 20th-century astrophysics. For the 21st century, the long-term goal is *to develop a comprehensive understanding of the formation, evolution, and destiny of the universe and its constituent galaxies, stars, and planets—including the Milky Way, the Sun, and Earth.*

In order to do this, the committee believes that astronomers must do the following:

- *Map the galaxies, gas, and dark matter in the universe, and survey the stars and planets in the Galaxy.* Such complete surveys will reveal, for example, the formation of galaxies in the early universe and their evolution to the present, the evolution of primordial gas from the Big Bang into matter enriched with all the elements by stars and supernovae, the formation of stars and planets from collapsing gas clouds, the variety and abundance of planetary systems in the Galaxy, and the distribution and nature of the dark matter that constitutes most of the matter in the universe.
- *Search for life beyond Earth, and, if it is found, determine its nature and its distribution in the Galaxy.* This goal is so challenging and of such importance that it could occupy astronomers for the foreseeable future. The search for evidence of life beyond Earth through remote observation is a major focus of the new interdisciplinary field of astrobiology.
- *Use the universe as a unique laboratory to test the known laws of physics in regimes that are not accessible on Earth and to search for new physics.* It is remarkable that the laws of physics developed on Earth appear to be consistent with phenomena occurring billions of light-years away and under conditions far more extreme than those for which the laws were derived and tested. However, researchers have only begun to probe the conditions near the event horizons of black holes or in the very early universe, where the tests of the laws of physics will be much more stringent and where new physical processes may be revealed that shed light on the unification of the forces and particles of nature.
- *Develop a conceptual framework that accounts for all that astronomers have observed.* As with all scientific theories, such a framework must be subject to continual checks by further observation.

For the new decade, astronomers are poised to make progress in five particular areas:

1. Determining the large-scale properties of the universe: its age, the nature (amount and distribution) of the matter and energy that make it up, and the history of its expansion;
2. Studying the dawn of the modern universe, when the first stars and galaxies formed;
3. Understanding the formation and evolution of black holes of all sizes;

TABLE 2.1 Science Goals for the New Initiatives

	Initiative[a]	
Science Goal	Primary[b]	Secondary[b]
Determining large-scale properties of the universe	NGST, GSMT, LSST (MAP, Planck, SIM)	Con-X
Studying the dawn of the modern universe	NGST, SKA, LOFAR (ALMA)	Con-X, EVLA, SAFIR, GLAST, LISA, EXIST, SPST
Understanding black holes	Con-X, GLAST, LISA, EXIST, ARISE	EVLA, LSST, VERITAS, SAFIR
Studying star formation and planets	NGST, GSMT, EVLA, LSST, TPF, SAFIR, TSIP, CARMA, SPST (ALMA, SIM, SIRTF, SOFIA)	AST, SDO, Con-X, EXIST
Understanding the effects of the astronomical environment on Earth	LSST, AST, SDO, FASR	GLAST

NOTE: Acronyms are defined in the appendix.

[a]Missions and facilities listed in parentheses are those that were recommended previously but have not yet begun operation.

[b]Projects or missions listed in the "primary" category are expected to make major contributions toward addressing the stated goal, while "secondary" projects or missions would have capabilities that address the goal to a lesser degree.

4. Studying the formation of stars and their planetary systems, and the birth and evolution of giant and terrestrial planets; and

5. Understanding the effects of the astronomical environment on Earth.

Table 2.1 lists these science goals and the new initiatives that will address them.

In addition, the time is ripe for *using astronomy as a gateway to enhance the public's understanding of science and as a catalyst to improve teachers' education in science and to advance interdisciplinary training of the technical work force.*

THE FORMATION AND EVOLUTION OF PLANETS

The discovery of extrasolar planets in the past decade was one of the most remarkable achievements of the 20th century and represented the culmination of centuries of speculation about planets orbiting stars other than our Sun. These observations confirmed for the first time that a significant fraction of the stars in the Milky Way Galaxy have planetary systems; at the same time, the observations brought the surprising news that a number of planetary systems are very different from our solar system. In fact, the first extrasolar planetary system discovered is quite exotic: Although it involves terrestrial-mass planets, the central star is not a normal star like the Sun, but a rapidly spinning neutron star. The first planet detected around a Sun-like star is much more massive than Earth. Its mass is at least half that of Jupiter, the largest planet in the solar system, but its orbit is only one-tenth as large as that of the innermost planet, Mercury (Figure 2.1). Further discoveries indicate that such "hot Jupiters"—gas giant planets orbiting 100 times closer to the host star than their analogs in our own solar system—are surprisingly common, being found around a few percent of all solar-type stars. It may even be that our own planetary system is the exception and hot Jupiters the rule.

We are witnessing the birth of a new observational science of planetary systems. The new measurements of masses and orbital distances of planets demand explanation. The first step is to carry out a census of extrasolar planetary systems in order to answer the following questions: What fraction of stars have planetary systems? How many planets are there in a typical system, and what are their masses and distances from the central star? How do these characteristics depend on the mass of the star, its age, and whether it has a binary companion?

Astronomers have a number of methods to detect extrasolar planets: astrometry, measurement of Doppler shifts, photometry, observations of gravitational microlensing, and direct imaging. SIM will utilize astrometry, a method that uses the back-and-forth motion of stars in the sky to infer the presence of an orbiting planet, to increase the census of Jovian-mass planets orbiting at relatively large distances from their central stars. GSMT and other ground-based telescopes will measure small shifts in the wavelengths of the observed radiation, or the Doppler shifts, caused by the motion of stars toward and away from us as the planets orbit the stars. The Doppler method has been used almost exclusively in the past decade and favors small orbital separation and

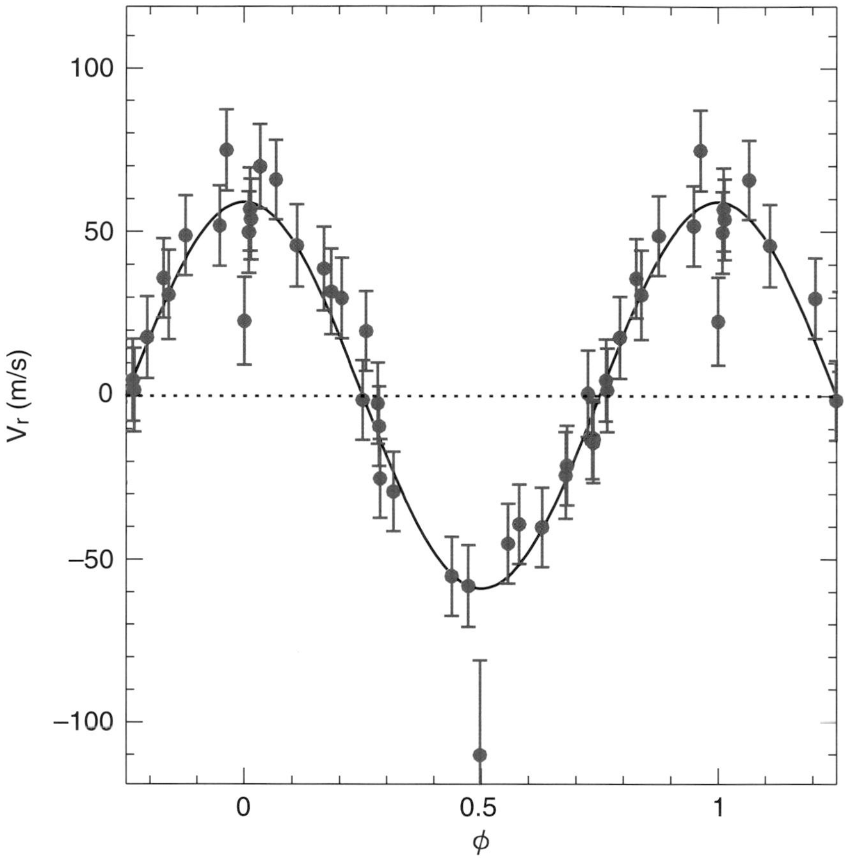

FIGURE 2.1 The discovery of the first planet orbiting a Sun-like star outside the solar system was made by observing small oscillations in the radial velocity V_r of the star 51 Pegasi. These oscillations are caused by the planet as it orbits the star every 4.2 days. The phase ϕ represents the time in units of the 4.2-day cycle. Courtesy of M. Mayor, D. Queloz, and S. Udry (Université de Genève). Reprinted by permission from *Nature* 378:355-359, copyright 1995 Macmillan Magazines Ltd.

relatively large planets. Photometry measures the small decrease in the light from a star when a planet orbits between the observer and the star, partially eclipsing the star. Because photometry depends on a favorable inclination of the orbit, surveys of a large number of stars are required to find the frequency of planetary systems. Space-based photometry is sufficiently precise that it could extend the census to planets with masses as low as those of the terrestrial planets. Sensitive photometry of distant stars can also reveal planets through gravitational microlensing: The gravitational field of an intervening faint star close to the line of sight to a distant star acts as a lens that amplifies the light of the distant star; planets orbiting the intervening star can change the amplification in a detectable manner. However, these methods all detect planets indirectly by their small perturbations of the light from the central star. The ultimate goal is to see and study the radiation from the planets themselves. Direct imaging of giant planets can be done from the ground with adaptive optics, but TPF or an enhanced NGST is needed for terrestrial planets.

Once direct imaging is possible, radiation from extrasolar planets can be analyzed to characterize the atmospheres of the planets: How do the atmospheres depend on the mass of the planet, its separation from its host star, and the mass of the host star? Do any of the planets appear habitable? Are there any biological "marker materials" such as methane, molecular oxygen, or ozone? Observation of the atmospheres is extremely challenging, owing to confusion with the enormously brighter host star. TPF is designed to address this problem by using interferometry to null out the radiation from the host star; with the addition of an occulter NGST may contribute to this goal.

The planetary census, together with new observations of protoplanetary disks, will provide the data needed to understand planet formation. Observations over the past two decades have established that protostars are accompanied by disks of gas and dust. These disks are believed to feed the growth of the stars and are regions where planets could form. Today's instruments do not have the resolution or the sensitivity to find evidence for the existence of planets in protostellar disks, but ALMA, NGST, and TPF will. Theory shows that gas giants should create gaps in the disks that will be readily observable by these powerful instruments. Young giant planets ($\lesssim 10$ million years old) will emit enough radiation in the near infrared to be detectable by both NGST and GSMT in the nearby molecular clouds where star formation is occurring. These observations will reveal how protostellar disks evolve and the conditions under which planets can form. The existing census of extrasolar planets already indicates a surprising number of massive planets orbiting extremely close to the central star. Are these planets formed in the outer regions of the disk and then pushed into tighter orbits by the gravitational interaction with the disk material or with other planets? The Sun is in the minority in not having a stellar companion. Now do companion stars affect planet formation? Most stars form in large clusters containing massive stars, such as the cluster associated with the Trapezium in Orion. What is the effect of such an environment on planet formation? Hubble pictures showing the destruction of protostellar disks in the Orion Nebula (Figure 2.2) suggest that such an environment is very hostile to planet formation.

Some recent discoveries within our own solar system point the way toward another approach to filling in some details of the picture of planet formation and evolution. The Kuiper Belt consists of a ring or disk of subplanetary bodies circling the Sun beyond Neptune. Some 200 Kuiper Belt objects (KBOs) are now known, with diameters mostly in the 100- to 800-km range (Figure 2.3). Smaller KBOs are too faint to have been detected in existing surveys; larger ones almost certainly exist but await detection by deep, all-sky surveys such as will be conducted by LSST. It is thought that as many as 10 more objects of Pluto size (with a diameter of 2,000 km) await discovery. These KBOs are but the tip of an iceberg. Probably 100,000 objects larger than 100 km exist at distances 30 to 50 times Earth's distance from the Sun. The number of objects larger than 1 km lies in the range of 1 billion to 10 billion. These objects are fossil remnants of the Sun's planetary accretion disk, and their motions provide direct evidence of the protoplanetary disk's physical characteristics. Collisions between these objects provide a long-term source for tiny dust particles in the solar system. Similar dust disks have been detected recently around some other main-sequence stars. The Kuiper Belt is probably the source of most short-period comets. Near-infrared spectra of the KBOs capitalizing on the huge light-collecting capability of GSMT will, for the first time, reveal the composition of comets in their pristine state, prior to entry into the inner solar system.

The atmospheres of planets can be studied primarily in our own solar system. Except for Uranus, the gas giant planets emit more energy than they receive from the Sun. Their internal heat production drives complex and poorly understood systems of convection. The main external manifestations include differential rotation (as in the Sun) and energetic, weather-like, circulation patterns at the visible cloud tops. Planetary convection also powers dynamo action, causing the gas giants to support huge radio-bright magnetospheres. New adaptive optics systems on large-aperture telescopes will provide 10-milliarcsec resolution in the near infrared (Figure 2.4), enabling the study of long-term changes in planetary circulation (at Jupiter, 10 milliarcsec = 35 km; at Neptune, 200 km). Such studies will also provide the context for in situ investigations by NASA spacecraft.

Near-Earth objects (NEOs) are asteroids with orbits that bring them close to Earth. The orbits of many NEOs actually cross that of Earth,

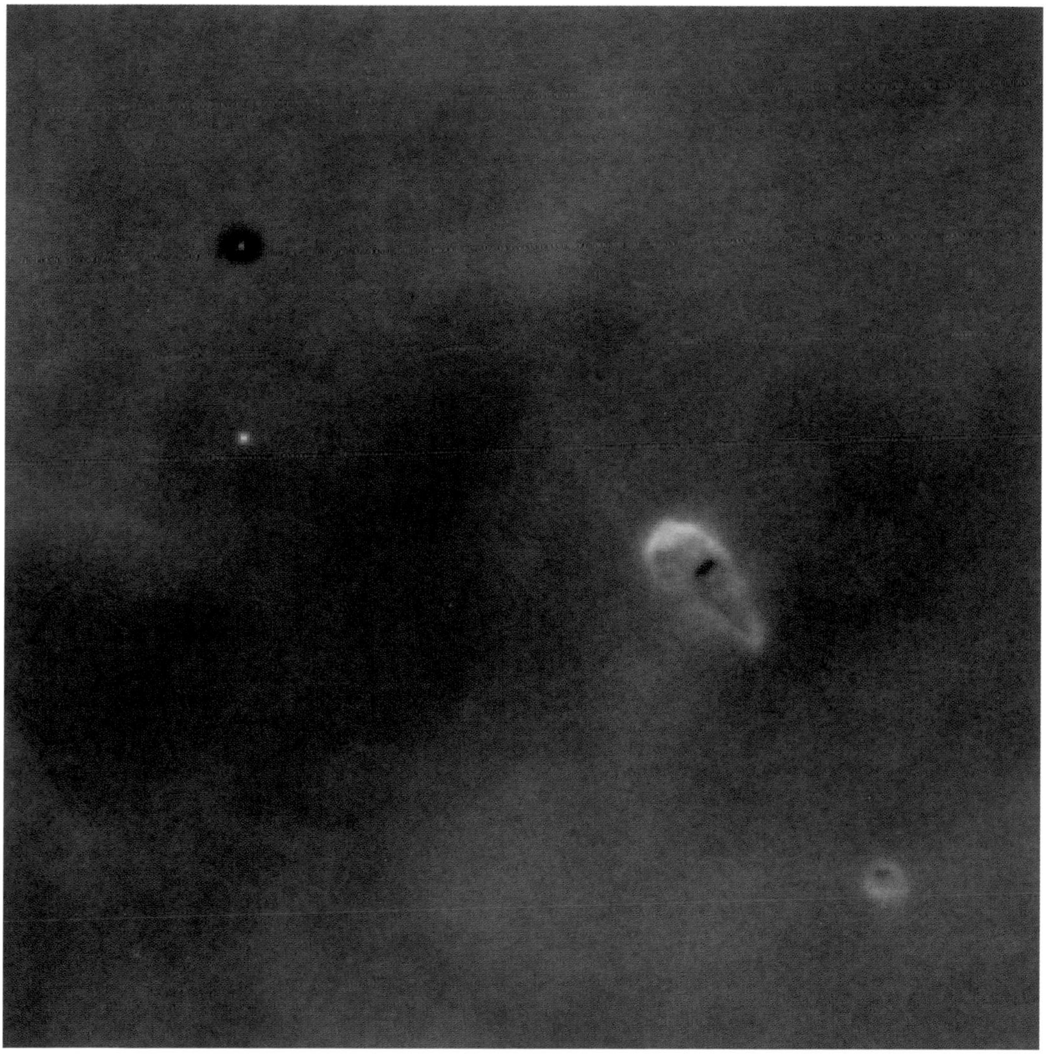

FIGURE 2.2 Protoplanetary disks in the Orion Nebula. These dark silhouetted disks, sometimes surrounded by bright ionized gas flows as seen in the cometary shape above, are being destroyed by intense ultraviolet radiation from nearby massive stars. The rapidity of their destruction may interrupt planet formation in these disks. Courtesy of C.R. O'Dell (Rice University) and NASA.

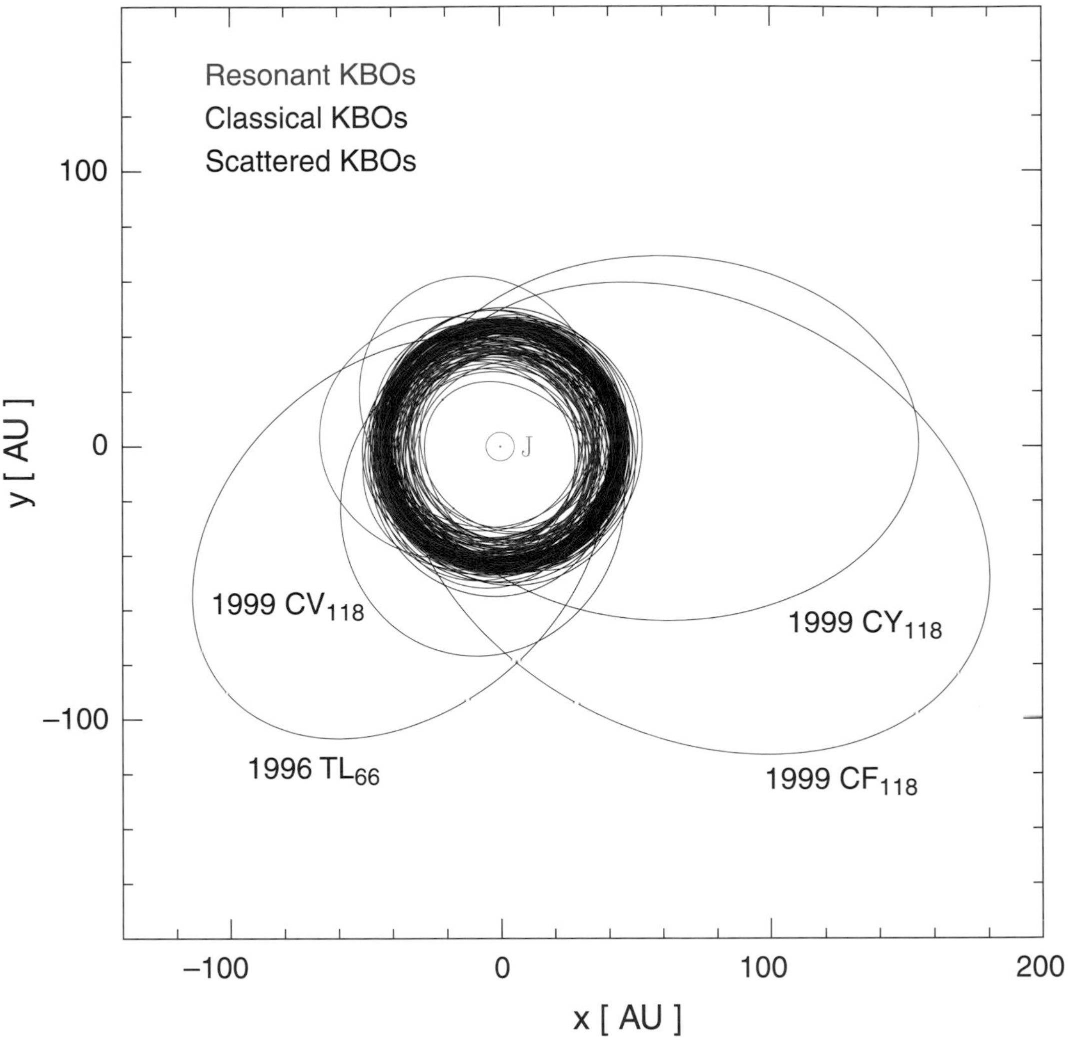

FIGURE 2.3 Plan view of the solar system, showing the orbits of the 200 Kuiper Belt objects (KBOs) known as of October 1999. Red orbits denote KBOs in orbits that are in resonance with Neptune, including Pluto; blue orbits show nonresonant or "classical" KBOs; and the large, eccentric orbits with labels denote KBOs that have been scattered by the gravity of the giant planets. The orbit of Jupiter at 5 AU (AU = astronomical unit, the distance from Earth to the Sun) is shown for scale. Observations with LSST should increase the number of known KBOs to 10,000, permitting intensive investigation of the dynamical structure imprinted on this fossil protoplanetary disk by the formation process. Courtesy of D. Jewitt (University of Hawaii).

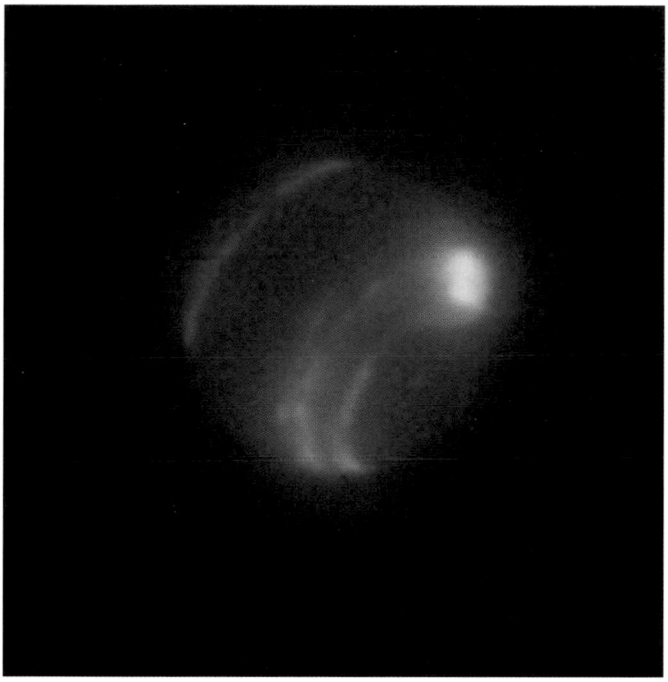

FIGURE 2.4 An image of Neptune taken by the Keck Adaptive Optics Facility in the methane absorption band at 1.17 μm. The angular resolution of this image is approximately 0.04 arcsec, about an order of magnitude better than the resolution obtained without adaptive optics. Courtesy of the W.M. Keck Observatory Adaptive Optics Team. (This figure originally appeared in *Publications of the Astronomical Society of the Pacific* [Wizinowich, P., et al., 2000, vol. 112, pp. 315-319], copyright 2000, Astronomical Society of the Pacific; reproduced with permission of the Editors.)

making NEOs an impact threat to our planet. Extrapolations from existing data suggest that about 1,000 NEOs are larger than 1 km in diameter, and that between 100,000 and 1 million are larger than 100 m. The effects of past NEO impacts on Earth range from the destruction of hundreds of square miles of Siberian forest at Tunguska in 1908 by a relatively small NEO to substantial disruption of the biosphere at the end of the Cretaceous period some 60 million years ago by a large (10-km) NEO. Interplanetary space is vast, so the probability of a substantial NEO hitting Earth is small: For example, it is estimated that the probability that an NEO larger than 300 m will strike Earth during this century is about 1 percent. Nonetheless, it behooves us to learn much more about these objects. Over a decade, LSST will discover 90 percent of the NEOs larger than 300 m, providing information about the origin of these objects in the process. However, comets also pose a substantial impact hazard, as was dramatically illustrated by the impact of Comet Shoemaker-Levy on Jupiter (Figure 2.5). Although LSST will discover much about comets, it will not provide long-term warning of potentially hazardous long-period comets.

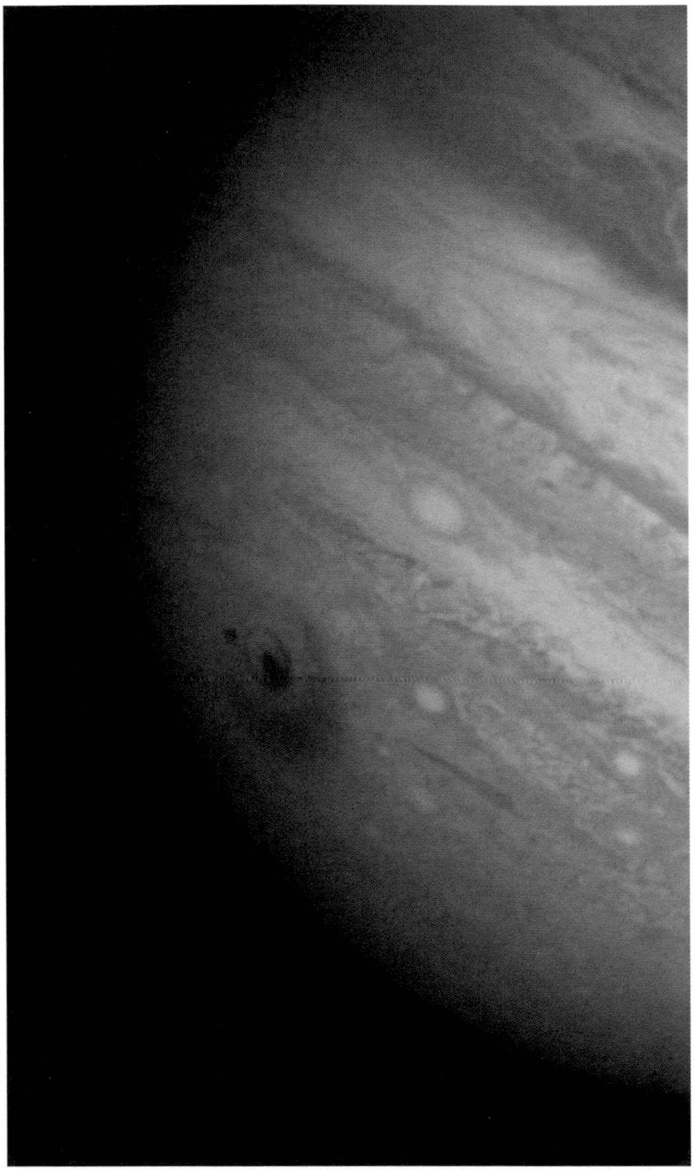

FIGURE 2.5 The impact of fragment "G" of Comet Shoemaker-Levy 9 onto Jupiter in July 1994 left dark rings of substantially altered atmosphere (lower left section of the planet). The thick dark outermost ring's inner edge has a diameter about the size of Earth's. The impact had an explosive energy equivalent to roughly a million megatons of TNT. Courtesy of H. Hammel (Massachusetts Institute of Technology) and NASA.

STARS AND STELLAR EVOLUTION

The development and confirmation of the theory of the structure and evolution of stars represent one of the great achievements of 20th-century science. Stars are the building blocks of galaxies and are the "atoms" of the universe. Essentially all the elements in our bodies except hydrogen were created in the nuclear fires in stellar interiors. The discovery in the past decade of "brown dwarfs," stars too small to burn hydrogen, has extended the range of stellar masses over which the theory applies. Despite the great success of this theory, it has a gaping hole: It neither predicts nor explains how stars form. Such knowledge is critical for understanding not only how planets form, but also how systems of stars, such as galaxies, must evolve.

STAR FORMATION

Star formation proceeds in the densest regions of opaque clouds of gas and dust that are scattered throughout the interstellar medium of a galaxy (Figure 2.6). Most of the gas in these clouds is molecular, and it is highly inhomogeneous. Stars form in the densest parts of molecular clouds when the mutual gravitational attraction of the gas overcomes the thermal pressure, turbulent motions, and magnetic fields that support the cloud. The ensuing collapse forms a single star, a binary, or less often, a multiple-star system. Theory suggests, and observations confirm, that most stars are encircled by disks when they first form. These disks are the birthplaces of planets. As stars grow by accretion of material from their disks, powerful bipolar winds are created perpendicular to the disks. These winds interact strongly with the infalling material and the natal molecular cloud. The mass of a star is the primary determinant of its characteristics over most of its life, yet researchers do not know what determines the star's birth mass. There are many other important unsolved problems in star formation as well, including understanding how molecular clouds form in the interstellar medium, how these clouds evolve to form protostellar cores, what tips the scales in favor of gravitational collapse, what determines when binaries form, how stars form in clusters, and how protostellar winds affect star formation.

From a theoretical perspective, studying star formation is challenging because it requires following the evolution of matter from the very tenuous gas in the interstellar medium, where densities are measured in the number of particles per cubic centimeter, to stellar interiors, where

FIGURE 2.6 Pillars of interstellar gas being eroded by radiation from massive stars in the Eagle Nebula, revealing low-mass stars in the process of formation. HST image courtesy of J. Hester and P. Scowen (Arizona State University), and NASA.

the densities are measured in grams per cubic centimeter—a trillion trillion times greater. Nevertheless, considerable progress has been made toward developing a theory, particularly for isolated stars with masses similar to that of the Sun. Numerical simulation on supercomputers is playing an important role in this effort. Theories of massive star formation are less advanced because of the strong interaction of the radiation from these luminous stars with the infalling gas and dust. The theory of star formation in clusters is similarly primitive because of the complicated interaction of the cores and protostellar winds in these regions.

From an observational perspective, star formation is challenging because dust obscures the regions of star formation, rendering them largely invisible to optical telescopes. Observation of the formation of massive stars is even more challenging since the sites of massive-star formation are rare and therefore on average more distant; furthermore, recent observations show that they are obscured by even more dust than are the regions of low-mass star formation. Infrared, submillimeter, millimeter, and hard x-ray radiation penetrate the obscuring dust; in addition, the gas and dust that form stars, disks, and planets radiate primarily at infrared and longer wavelengths. The substantial improvements in sensitivity and spatial resolution at these wavelengths obtained with many of the recommended new initiatives, together with facilities now under development, should lead to great advances in solving the important problems in star formation (see Table 2.1).

THE SUN

As the nearest star, the Sun provides us with the opportunity to test with exquisite accuracy our understanding of stellar structure. Using a powerful combination of theory and observation, solar physicists have done just that over the past decade: By studying tiny oscillations in the Sun (a technique termed "helioseismology"), they have shown that theoretical models for the internal structure of the Sun are accurate to within about 0.1 percent. Solar models are sufficiently accurate that the Sun can now be used as a well-calibrated source of neutrinos to carry out investigations of the basic physics of these fundamental particles.

Although understanding of the equilibrium properties of the Sun has been validated by helioseismology, understanding of the nonequilibrium properties—associated primarily with magnetic fields—remains poor. Magnetic fields play a crucial role in astrophysical phenomena ranging from the formation of stars to the extraction of energy from supermassive black holes in galactic nuclei. The Sun provides a natural laboratory for the study of cosmic magnetism on scales not accessible on Earth and not resolvable in distant astronomical objects (see Figure 2.7). Solar magnetic fields lead to "space weather," which can destroy satellite electronics and disrupt radio communications. These fields are also believed to be responsible for the variations in the Sun's luminosity that lead to variations in Earth's climate on a time scale of centuries. Such climate variations have undoubtedly influenced the evolution of life on Earth. Other stars are observed to have larger variations in their luminosity, which could have a correspondingly stronger effect on any life that might exist on planets in those systems.

The first scientific goal for advancing the current understanding of solar magnetism is to measure the structure and dynamics of the magnetic field at the solar surface down to its fundamental length scale. This length scale is believed to be determined by the pressure scale height, which is about 70 km, or 0.1 arcsec in angle from Earth; numerical simulations suggest that the size of magnetic flux tubes might be about half this. AST is designed to achieve this angular resolution. With the collecting area of a 4-m mirror, it will also have sufficient sensitivity to measure weak magnetic fields on this scale at the requisite time resolution. AST will permit substantial progress in the understanding of the physical processes in sunspots. At night, AST will obtain complementary information on the role of stellar magnetic fields by observing other stars, which can behave quite differently from the Sun. Constellation-X will contribute to these studies by providing accurate measurements of physical conditions in the coronae of other stars.

The second scientific goal is to measure the properties of the magnetic field throughout the entire solar volume, extending from below the surface out to 18 solar radii. Below the visible surface of the Sun, magnetic fields are trapped in the solar gas and move with it. The turbulent convection and the apparently random emergence of magnetic fields cause surface magnetic fields to be mixed on a range of scales. An important development of the past decade was the use of acoustic tomography to create three-dimensional maps of these field structures. Above the surface, in the solar corona, the gas density drops very rapidly and the situation is reversed: There, the highly conducting solar gases are forced to move with the magnetic fields, so that the entire outer atmosphere responds continuously to the motions of the footpoints of the magnetic field trapped in the surface. Extreme ultraviolet measurements

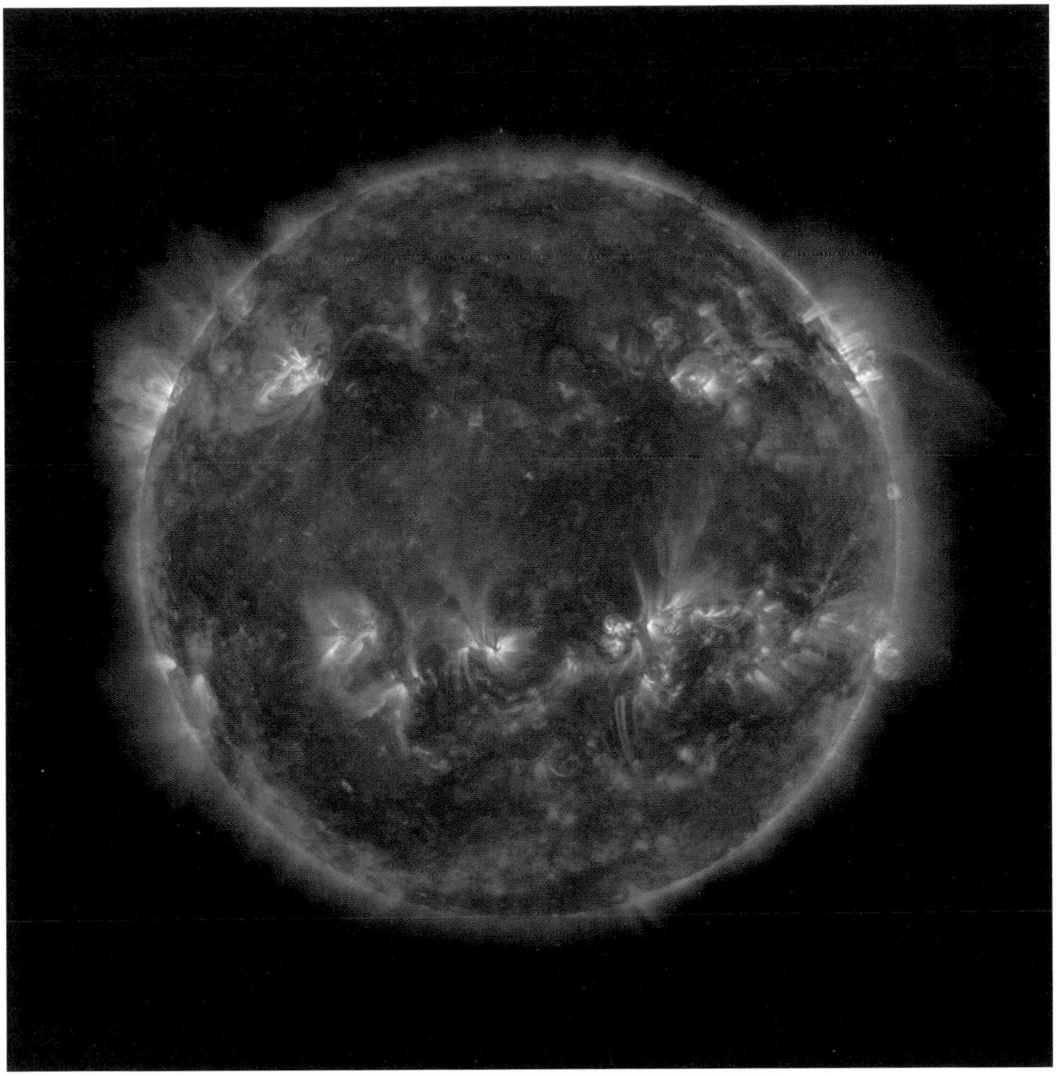

FIGURE 2.7 An image of the full disk of the Sun at x-ray wavelengths (0.0171 μm), which are sensitive to the emission from a highly ionized iron atom (eight electrons removed). This emission arises from gases with temperatures between 600,000 and 1 million K. The image is a photomosaic of 42 overlapping 8.5 by 8.5 arcsec images taken by the TRACE spacecraft. Courtesy of NASA and the Stanford-Lockheed Institute for Space Research.

made by the TRACE spacecraft have shown that as a result, coronal structures are rapidly evolving and highly inhomogeneous, with loops at 30,000 K adjacent to loops at 3 million K (see Figure 2.7). When regions with opposite polarity collide, the overlying magnetic fields reconnect and restructure. These processes release enormous amounts of energy that are responsible for the heating of the outer solar atmosphere, flares, coronal mass ejections, and the acceleration of the solar wind toward Earth. SDO, which combines observations of the subsurface, surface, and corona, is designed to collect data to answer fundamental questions about the interaction of gas flows and magnetic fields, reconnection and restructuring of magnetic fields, rapid energy release processes, and outward acceleration of solar material.

Together, AST and SDO will provide a comprehensive view of the dynamics of the solar magnetic field and lead to a much deeper understanding of cosmic magnetism. In addition, these projects will revolutionize our understanding of space weather and global change, which are influenced by the Sun because Earth and the space surrounding it are bathed by the Sun's outer atmosphere.

STELLAR METAMORPHOSIS

Most living things slow down as they age and eventually cease to be able to generate new life. Stars behave in the opposite fashion: Evolution accelerates when they near the end of their lives as normal stars, and during the final stages a significant fraction of their mass, enriched with heavy elements generated in their interiors, is dispersed into surrounding space (Figure 2.8). The ejected gas, mixed with the local interstellar medium, can then be recycled to form new stars and planetary systems. Left behind is a compact stellar remnant—a white dwarf, with a radius 100 times smaller than that of the Sun; a neutron star, with a radius 1,000 times smaller; or a black hole, with an effective radius that, for a mass comparable to that of a neutron star, is several times smaller yet. Stellar "death" is thus a metamorphosis in which stars that are powered by nuclear reactions, like the Sun, are reborn as compact objects.

Most stars with a mass more than about eight times that of the Sun end their lives in a titanic explosion, a supernova, leaving behind a neutron star or a black hole (Figure 2.9). Stars less massive than about eight times the mass of the Sun evolve into red giants, so large that at the position of the Sun they would envelop the orbit of Earth. Their distended envelopes are ejected soon afterward, leaving behind a white

FIGURE 2.8 Hubble Space Telescope image of the planetary nebula NGC 6543, commonly known as the Cat's Eye Nebula. The inset shows the lower-magnification ground-based image made using the 2.1-m telescope at Kitt Peak National Observatory under excellent atmospheric conditions. Stars with a mass less than about eight times that of the Sun evolve to red giant stars, and the red giants end their lives by ejecting their outer envelopes. The ejected envelopes glow in visible light and are called planetary nebulae. This image shows the ejected gas, enriched in elements such as carbon by the nucleosynthesis that occurred in the parent star, as it travels outward into the interstellar medium to be incorporated eventually into new stars and planets. The Hubble image was obtained by J.P. Harrington and K.J. Borkowski (University of Maryland), and NASA, and was recolored by B. Balick (University of Washington) with permission. The ground-based image is courtesy of B. Balick.

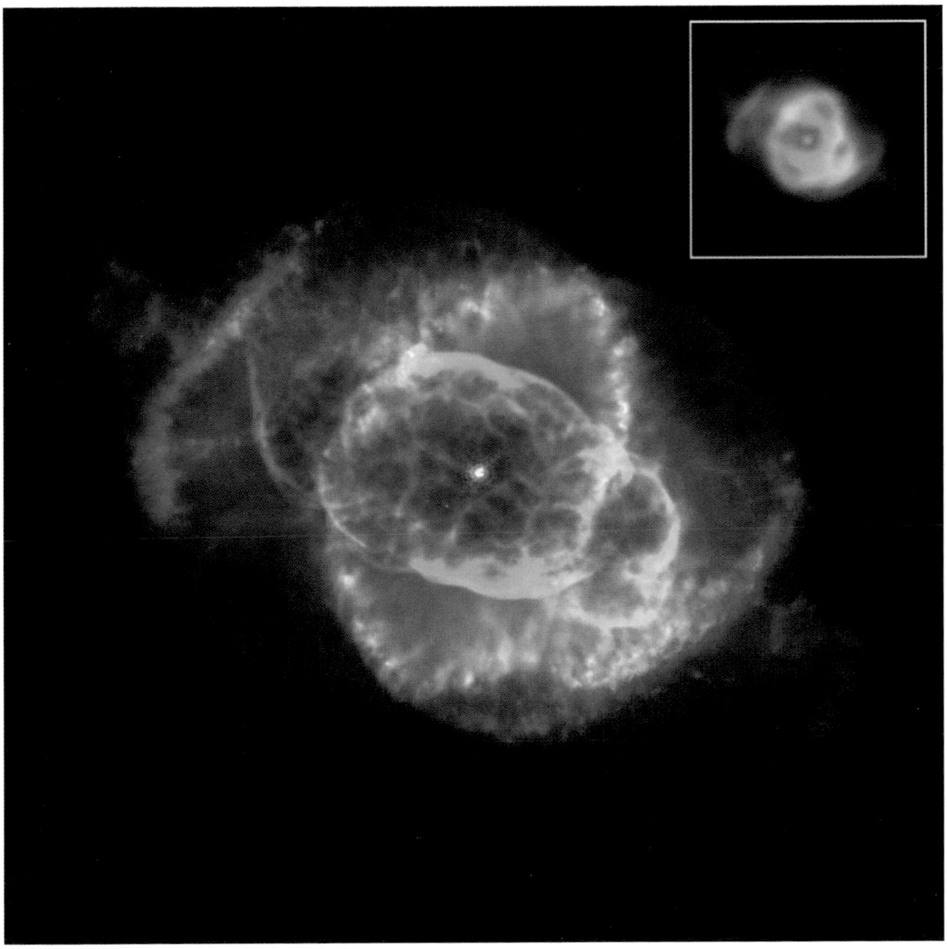

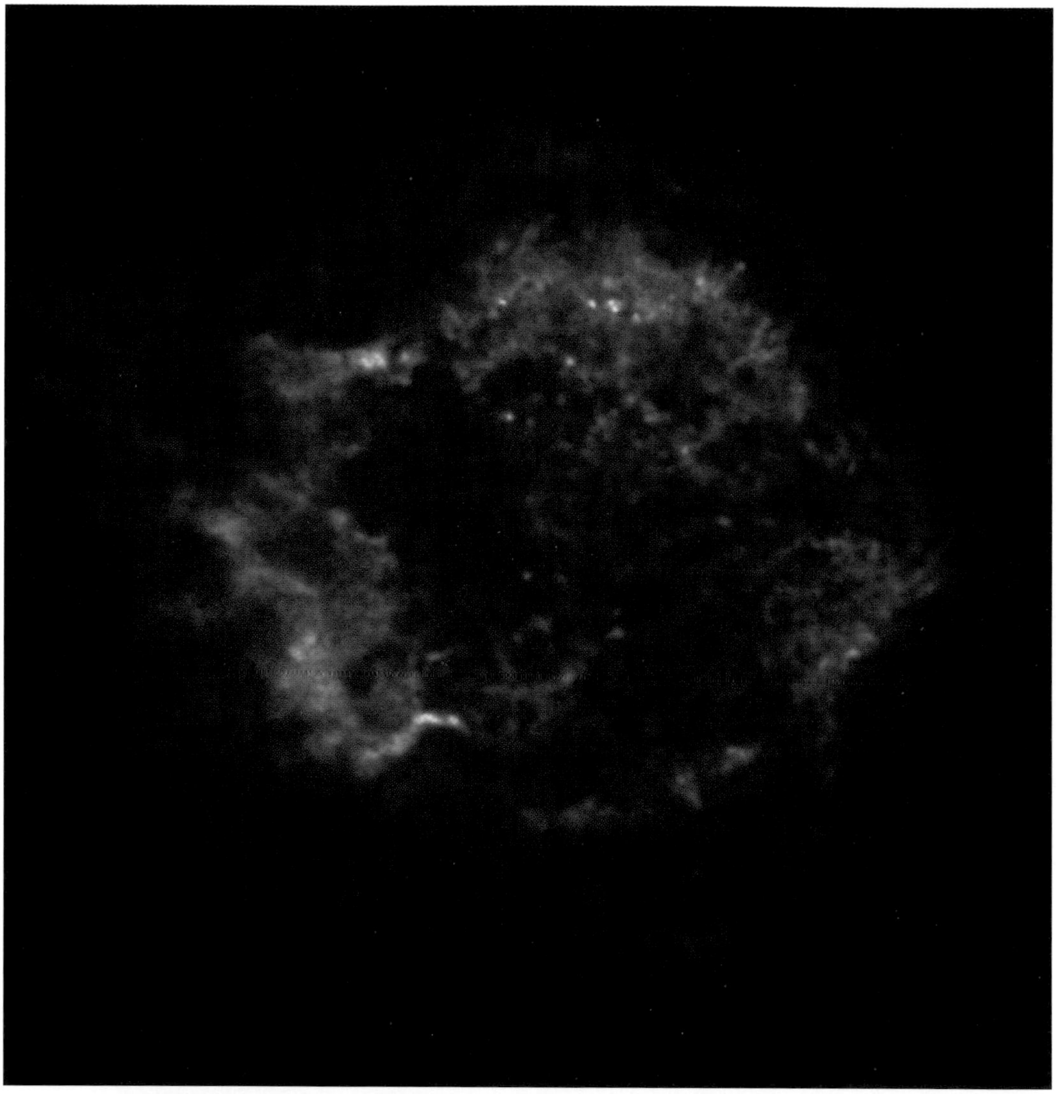

FIGURE 2.9 Two supernova remnants observed by the Chandra X-ray Observatory. On the left is an x-ray color image of Cassiopeia A, the remnant of a supernova that exploded about 300 years ago. The red, green, and blue regions show where the intensity of low-, medium-, and high-energy x rays, respectively, is greatest. The x rays from Cassiopeia A are produced by collisions between hot electrons and ions. The point source near the center is believed to be the compact

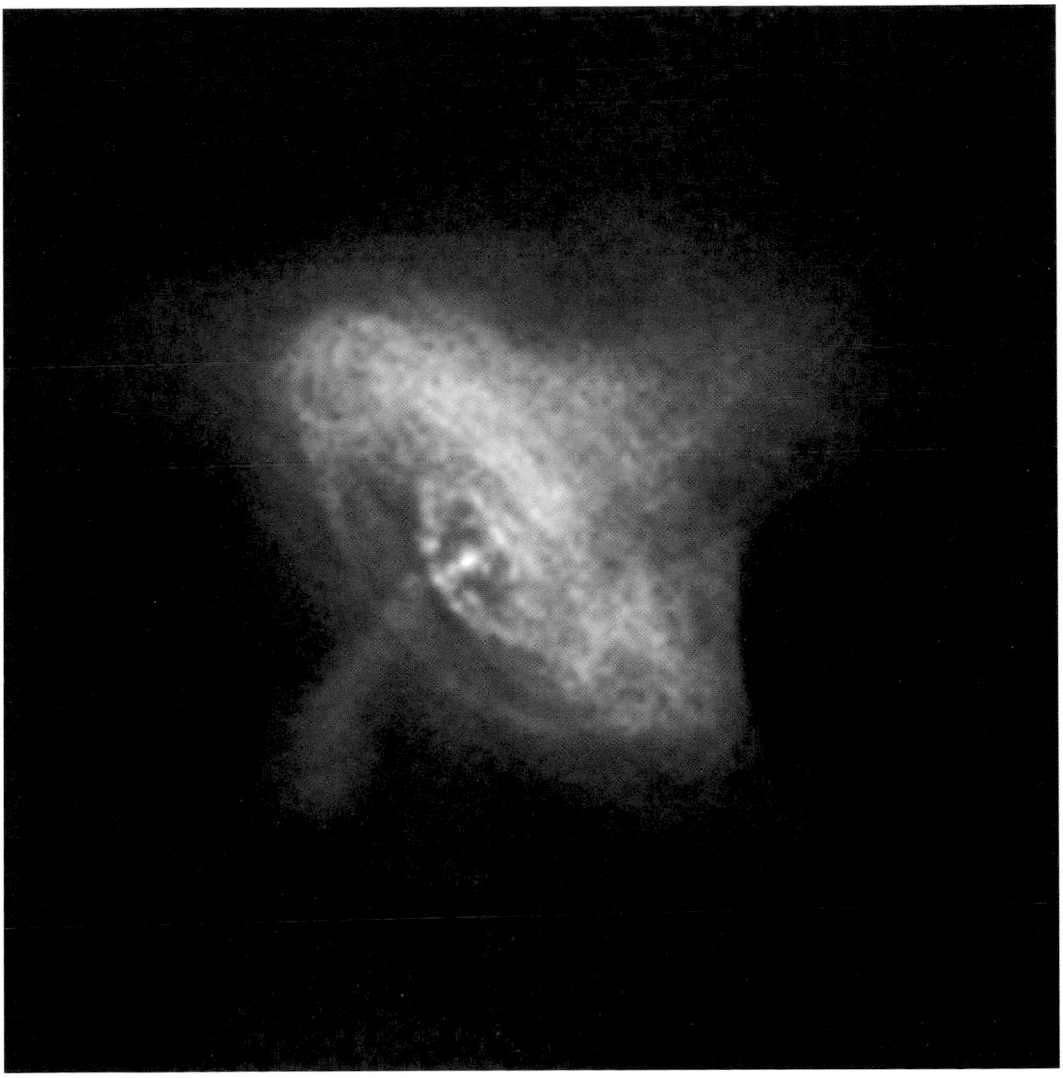

stellar remnant—a neutron star or black hole—left behind by the explosion. On the right is an x-ray intensity image of the Crab Nebula, the remnant of the supernova of 1054 AD. The x rays from the Crab Nebula are produced by electrons that accelerate to nearly the speed of light and then spiral in the magnetic field of the nebula. Images courtesy of NASA, the Chandra X-ray Observatory Center, the Smithsonian Astrophysical Observatory, and J. Hughes (Rutgers University).

dwarf remnant (see Figure 2.8). For all these stars, some newly created elements are ejected from the surface in stellar winds before the final collapse. A major goal of stellar astrophysics is to understand the various mechanisms of mass loss and how they contribute to the continually increasing abundance of heavier elements in the universe. Many of the recommended new facilities will make strong contributions to the necessary investigations: ALMA and CARMA by studying the chemistry of the outflows, GSMT by acquiring spatially resolved spectra, and Constellation-X by observing the newly formed elements in supernova ejecta.

If a white dwarf has a closely orbiting companion star, it may accrete matter from the companion and become a supernova itself. Such supernovae (called Type Ia) have luminosities that can be calibrated, so that they can be used as standard candles. This means that their apparent brightness can be converted to distance. By measuring the distances and redshifts of many supernovae, it is possible to probe the geometry of the universe (is it flat or curved?) and determine how its expansion rate is changing with time. One of the major goals of stellar research during this decade will be to understand Type Ia supernovae both observationally and theoretically in order to calibrate their luminosities. LSST will aid in discovering large numbers of supernovae, and both NGST and GSMT will enable detailed study of their spectra even when they are at high redshifts.

Stars that are reborn as compact objects have such strong gravitational fields at their surfaces that they radiate high-energy photons when material falls on them, thus making them observable in the x-ray region of the spectrum. Neutron stars and white dwarfs also radiate the thermal energy stored in them at birth, and if they are magnetized and spinning, they can accelerate particles that also radiate. These objects provide laboratories in which matter can be studied under extreme conditions that cannot be duplicated on Earth. For example, the past decade saw the discovery of the theoretically predicted "magnetars," which are neutron stars with magnetic fields 100 times that of normal neutron stars and a billion times that of the largest static fields in the laboratory. One of the major goals of Constellation-X is to image gas indirectly as it accretes onto a black hole, by studying how its spectrum evolves with time. Another goal is to measure accurately how the radius of a neutron star depends on its mass, which will tell researchers about the properties of matter at nuclear densities.

Gamma-ray bursts are mysterious phenomena discovered by satellites that were monitoring the skies for possible thermonuclear test explosions. At its peak, the energy flux observed from a single burst can be greater than that from all of the nighttime stars and galaxies in the universe! The apparent brightness of the bursts led many astronomers to conclude that they had to be in our galaxy, but during the 1990s the Compton Gamma Ray Observatory found them to be equally distributed over the whole sky and therefore almost certainly extragalactic. The Italian-Dutch BeppoSAX satellite permitted more accurate localization of a few of the bursts, leading to the discovery of theoretically predicted afterglows at other wavelengths (Figure 2.10). Observational monitoring of these afterglows confirmed that the bursts originate from the far reaches of the universe. While the precise origin of the bursts remains a mystery, it is believed that they are most likely associated with the formation of compact stellar objects such as neutron stars and black holes (Figure 2.11). With GLAST, EXIST, and MIDEX missions such as Swift, it will be possible to find gamma-ray bursts that are fainter than those previously visible and to locate them more quickly for prompt follow-up observations at other wavelengths. Because they are so luminous, bursts associated with the first generation of star formation may be detectable.

GALAXIES

On very large scales, galaxies are the building blocks of the universe, as fundamental to astrophysics as ecosystems are to the environment. They come in a variety of types, ranging from disk galaxies like the Milky Way to elliptical and irregular systems. While visible primarily through the light from the stars they contain, galaxies are actually far more complex than a simple grouping of stars. Most of their matter is "dark" in that it is not visible at the sensitivity limits of today's telescopes. Many galaxies, including our own, harbor supermassive black holes in their nuclei, and these will almost certainly have an important role in galactic evolution. Finally, in most galaxies there is a significant amount of gas and dust between the stars, out of which new stars continue to form.

FORMATION AND EVOLUTION OF GALAXIES

During the past decade astronomers were for the first time able to study galaxies so distant that their light was emitted when the universe was only a small fraction of its present age. From the work of Edwin

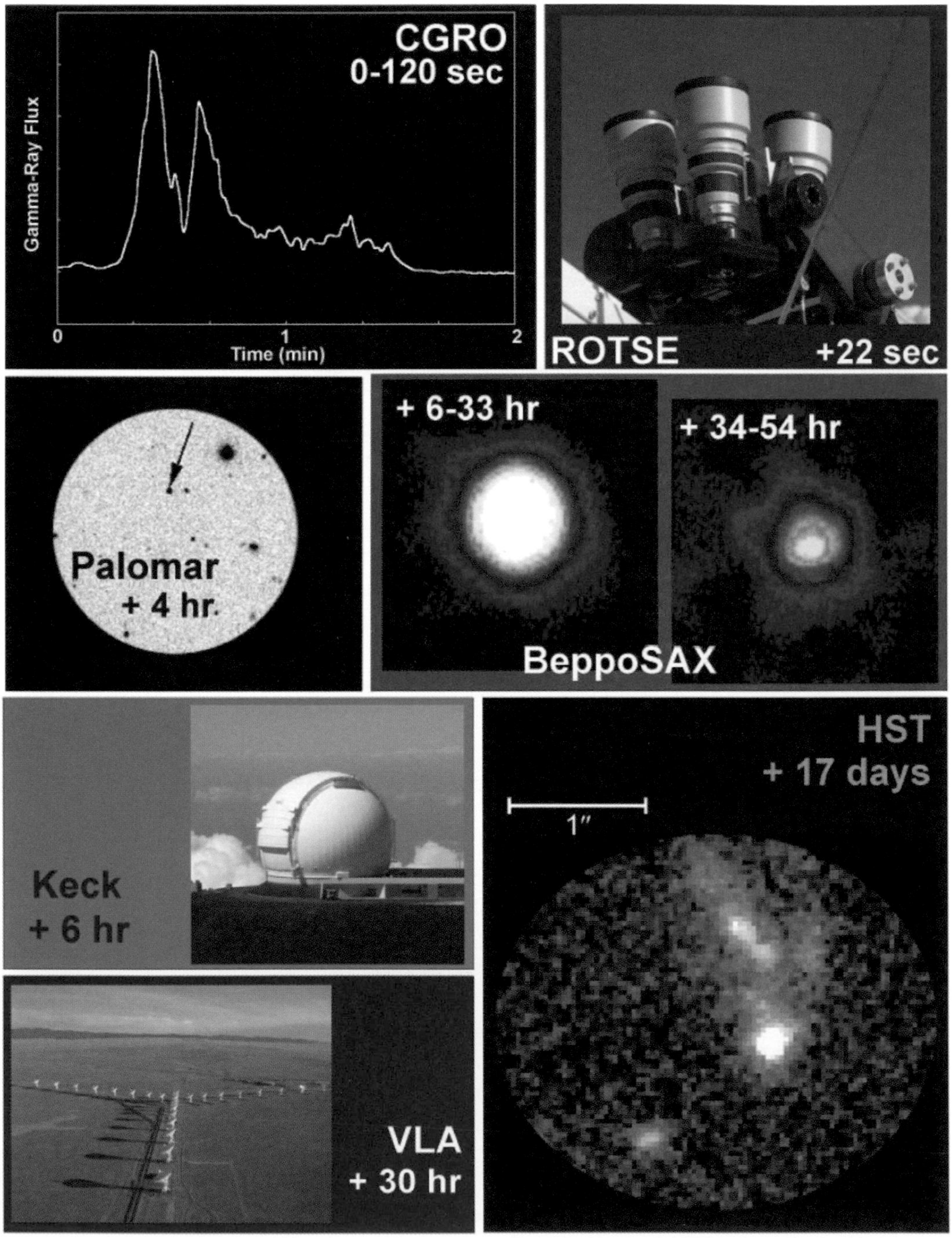
CGRO
0-120 sec
Gamma-Ray Flux
0
1
2
Time (min)
ROTSE
+22 sec
Palomar
+ 4 hr.
+ 6-33 hr
+ 34-54 hr
BeppoSAX
Keck
+ 6 hr
VLA
+ 30 hr
HST
+ 17 days
1″

FIGURE 2.10 Observations at many wavelengths are needed to understand gamma-ray bursts. This gamma-ray burst was discovered by the Compton Gamma Ray Observatory (CGRO) on January 23, 1999. The optical flash from the gamma-ray burst was observed by the Robotic Optical Transient Search Experiment (ROTSE) 22 seconds later. Subsequently, the BeppoSAX satellite detected the x-ray emission from the burst. Based on preliminary information from BeppoSAX, astronomers at the Palomar Observatory identified the precise location. Astronomers at one of the Keck telescopes were then able to obtain the spectrum and determine the distance. Within a day, radio astronomers used the Very Large Array to observe the fading afterglow of the burst. After 17 days, the burst had faded enough so that astronomers using the Hubble Space Telescope could observe the host galaxy. This is probably the most energetic gamma-ray burst ever recorded. Images courtesy of NASA, CGRO BATSE Team, ROTSE Project, J. Bloom (Caltech), BeppoSAX GRB Team, W.M. Keck Observatory, NSF/NRAO, A. Fruchter (STScI), and P. Tyler (NASA GSFC).

Hubble in the 1920s, astronomers have learned that the universe is expanding in such a way that distant galaxies are moving away from us at higher speeds than are nearby ones. The expansion of the universe "redshifts" radiation to longer wavelengths, or from blue to red. Greater redshifts correspond to more distant galaxies. Since it takes light longer to travel greater distances, greater redshifts also correspond to earlier epochs in the universe (Figure 2.12). Galaxies have been discovered at redshifts up to about 5.

Astronomers are also able to study galaxies at high redshifts by taking advantage of the sensitivity and angular resolution available with the Hubble Space Telescope (HST). Deep observations of two patches of sky, one in the north and one in the south, have revealed the morphology of these distant galaxies (the northern deep field is shown on the cover of this report). The conclusion of these studies is that galaxies have undergone enormous evolution since they were young, with large galaxies probably growing out of mergers of smaller ones. Observations at submillimeter wavelengths have suggested that some galaxies contain sufficient dust so that they reprocess a significant fraction of their starlight into far-infrared emission. As a consequence, optical and near-infrared observations are blind to as much as one-half of the star formation that has occurred in galaxies—a problem that observations with ALMA, SAFIR, FIRST, and SIRTF will overcome. SPST will survey the sky at submillimeter wavelengths, finding many high-redshift galaxies that these other telescopes can target.

Galaxies are often found in clusters, and these clusters are thought to grow in size by the merging of smaller clusters. As gas falls into clusters, it is heated to very high temperatures and emits x rays. Constellation-X will

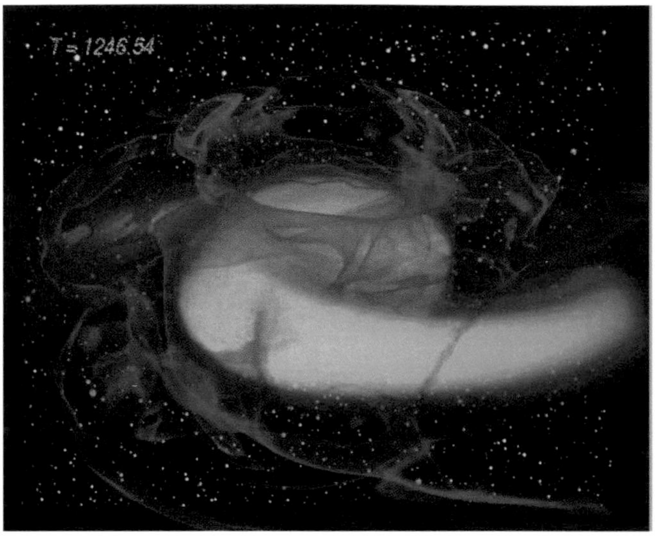

FIGURE 2.11 Simulation of two neutron stars spiraling into each other, ejecting hot gas (orange filamentary structures) and neutron-rich matter (blue/green, snail-shaped structure) in the process. Green represents higher-density matter than does blue. Gamma-ray bursts could be produced by such mergers. The white dots represent background stars added for visual effect. Simulation by P. Gressman (Washington University in St. Louis), and visualization by W. Benger (Max-Planck-Institut für Gravitations Physik, Konrad-Zuse-Institut). Courtesy of the NASA Neutron Star Grand Challenge Project.

be able to observe this emission from the first clusters of galaxies that form in the universe, revealing how they formed. Complementary observations with NGST and GSMT will show the evolution of clustering in cosmic time and how the cluster environment affects the evolution of galaxies.

As remarked above, present observations of galaxies do not extend much beyond a redshift of 5. The time between the "recombination" epoch at a redshift of about 1,000, when the cosmic background radiation was emitted, and that of redshift 5 remains completely unexplored. This period contains the "dark ages," when the visible light of the Big Bang faded and darkness descended. The dark ages ended with the formation of the first stars and galaxies—the dawn of the modern universe. The new decade brings the possibility of seeing the first generation of stars and galaxies that mark this dawn. NGST is designed to have the sensitivity and wavelength coverage to detect light from the first generation of galaxies, out to a redshift of about 20. With NGST it will be possible to address a number of fundamental questions: When did the first galaxies and stars form? What is the history of galaxy formation in the universe? What is the history of star formation and element production in galaxies? The ability of ground-based optical and infrared telescopes to address these questions is severely compromised by the opacity and the thermal emission from the atmosphere at wavelengths longer than 2 μm. NGST will cover the spectrum out to wavelengths of at least

5 μm, so that, for example, it can observe the hydrogen-alpha line produced in regions of massive star formation to a redshift of about 6 and the 0.4-μm stellar absorption feature to a redshift in excess of 10. Extending the sensitivity of NGST farther into the thermal infrared would greatly increase its ability to study galaxies at high redshifts.

Most of the stars and most of the heavy elements in the universe were formed after the epoch corresponding to redshift 5. As described above, the past decade has seen pioneering studies of galaxies in this redshift range, but the sensitivity and resolution have not been adequate to determine how the morphological and dynamical structure of galaxies has evolved over time. With adaptive optics and its enormous light-gathering power, GSMT will be a powerful complement to NGST for addressing such questions. Existing observations indicate very disturbed morphologies, possibly due to mergers, for galaxies at redshifts beyond 1;

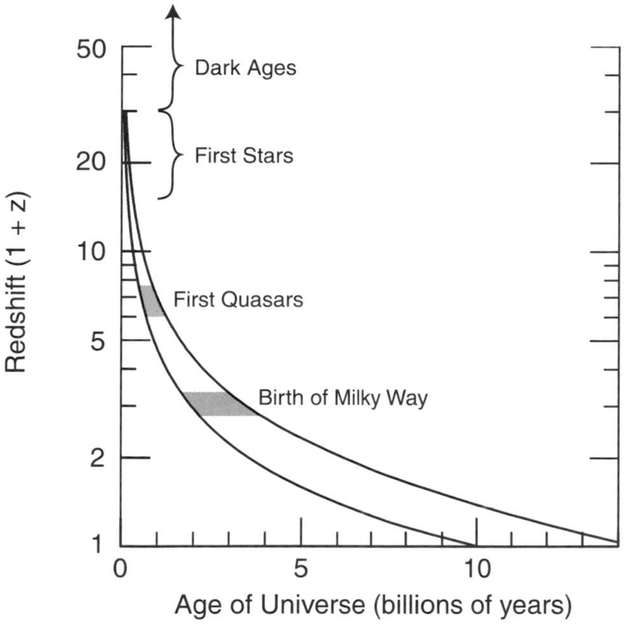

FIGURE 2.12 Relationship between the redshift and the time since the Big Bang (the age of the universe) for two different cosmological models. Both models are flat, but the one represented by the bottom curve has the critical density in matter, whereas that represented by the top curve (the currently favored model) has a cosmological constant so that only 30 percent of the critical density is in matter. At the time of the Big Bang, the age of the universe was zero, and the redshift (z) was extremely large. Later, prior to the formation of the first stars, the universe went through an epoch in which there was very little optical light—the "dark ages." Estimates are shown of the redshifts at which the first stars formed, quasars formed, and the Milky Way formed. Today, $1 + z = 1$; the age of the universe is 14 billion years in the currently favored model. See discussion on p. 86. Courtesy of M. Turner (University of Chicago).

GSMT and NGST will be able to distinguish the effects of mergers from those of rapid star formation. By means of spatially resolved spectroscopy, GSMT will be able to measure the masses of distant galaxies, thus providing crucial data for studying how galaxies evolve.

The history of galaxy evolution can also be inferred by studying the stellar populations of local galaxies at the present epoch. To do this requires the determination of the ages and elemental abundances of stars as a function of position in nearby galaxies. The high angular resolution available with GSMT means that it will be able to obtain the spectra of individual stars close to the nuclei of the Milky Way's nearest large companion galaxies, M31 and M32 (Figure 2.13).

EVOLUTION OF THE INTERSTELLAR MEDIUM IN GALAXIES

The interstellar medium in a galaxy controls the rate of star formation and thus the evolution of the galaxy itself. It is the repository of the heavy elements produced in stars. If star formation becomes too violent, interstellar gas may be ejected from a galaxy into the surrounding intergalactic medium. An understanding of the interstellar medium is necessary if researchers are to address such key questions as the following: What are the physical processes that determine the rate at which stars form in a galaxy? What is the feedback between star formation and the interstellar medium? (See Figure 2.6 for an example.) What is the effect of the extragalactic environment on star formation?

All these issues come into play when the formation of the first galaxies is considered. The first galaxies formed out of enormous clouds of neutral atomic hydrogen. Once the galaxies had formed, the interstellar media of these galaxies remained primarily atomic hydrogen, although with increasing amounts of heavier elements as massive, short-lived stars ejected new elements into the medium. The hydrogen gas should be observable at redshifts above 10 with LOFAR. When the SKA is built, it will be able to map the atomic hydrogen up to redshifts of about 10. Within galaxies, some of the atomic gas will be converted to molecular form on its way to being incorporated into stars. If the earliest stars have ejected enough carbon and oxygen into the interstellar medium, the broad spectral capabilities of the EVLA will enable observation of carbon monoxide, the most abundant molecule after molecular hydrogen, out to redshifts beyond 10. Newly formed stars ionize some of the gas, producing emission lines detectable by NGST. Supernovae heat large volumes

FIGURE 2.13 This optical wavelength picture shows the large spiral galaxy M31 (also known as the Andromeda Galaxy) and its small companions M32, lower center, and M110, to the upper right. Andromeda is the Milky Way's closest large neighbor at a distance of about 2.2 million light-years, and it is very similar in appearance to, and slightly larger than, the Milky Way. In fact, M31 is visible to the naked eye, although we can see only the bright inner bulge. This image comes from photographic plates taken with the 0.6-m Burrell Schmidt telescope of the Warner and Swasey Observatory of Case Western Reserve University. GSMT will be able to study individual stars near Andromeda's center, which is a very tightly packed star cluster not visible in this saturated image. Courtesy of B. Schoening (National Optical Astronomy Observatories) and V. Harvey (University of Nevada, Las Vegas, Research Experience for Undergraduates program sponsored by AURA/NOAO/NSF).

of the interstellar gas to millions of degrees, and x rays from this hot gas will be measured by Constellation-X to determine the temperature, pressure, and elemental abundances in this hot plasma. These same instruments will also permit astronomers to trace the evolution of gas in galaxies through cosmic time, as the universe synthesizes the elements needed to form planets and eventually to enable life.

Structure in the interstellar medium of a galaxy spans a wide range of scales, from much less than 1 light-year for the molecular cores that produce individual stars to 100,000 light-years for the galaxy as a whole. The gaseous galactic halo extends farther; it comprises both gas blown out of the disk and gas accreting from the intergalactic medium. Much of the mass of interstellar gas in disk galaxies is atomic and molecular gas that is quite cold, with a temperature that is less than 100 degrees above absolute zero. A substantial (but uncertain) fraction of the volume of such galaxies is filled by gas that has been heated to more than a million degrees by supernova explosions. There is also a significant amount of gas at intermediate temperatures that is heated by starlight. All this gas is permeated by cosmic rays, particles moving almost at the speed of light, and by magnetic fields. The primary hindrance to a greater understanding of how the interstellar medium mediates the evolution of galaxies is ignorance of the spatial distribution of these various components of the interstellar medium and how they are interrelated. Surveys of the interstellar medium in nearby galaxies with the recommended radio, infrared, x-ray, and gamma-ray facilities will provide valuable data on these issues. Understanding the complex structure of the interstellar medium and how it interacts with the process of star formation is a daunting theoretical problem for this decade.

GALACTIC NUCLEI

The nucleus of a galaxy is like a deep well: It is easy to fall in, but hard to get out. As a result, gas and stars accumulate there. In the 1960s, astronomers discovered that some galactic nuclei were truly remarkable: They could outshine an entire galaxy from a volume not much larger than that of the solar system. These objects, termed quasars, are the most luminous type of active galactic nucleus. Theorists immediately conjectured that such prodigious power output could come only from the accretion of gas onto a supermassive black hole; later it was realized that energy could be extracted from the spin of the black hole as well. A consequence of these ideas is that many galaxies should harbor supermassive black holes in their nuclei. Three decades later, this conjecture has been amply verified. Observations of both gas and stars have shown that even in our own "backyard," the Milky Way Galaxy harbors a black hole 3 million times more massive than the Sun (Figure 2.14)—and that black hole masses in the nuclei of other galaxies can exceed a billion solar masses. Exquisitely precise measurements of the positions and three-dimensional velocities of water masers made with the Very Long Baseline Array (VLBA) toward the nucleus of the galaxy NGC 4258 provided incontrovertible evidence for the presence of a supermassive black hole (Figure 2.15). ARISE has the power to study the water emission in other galactic nuclei to search for black holes and determine their mass and the characteristics of the accreting gas.

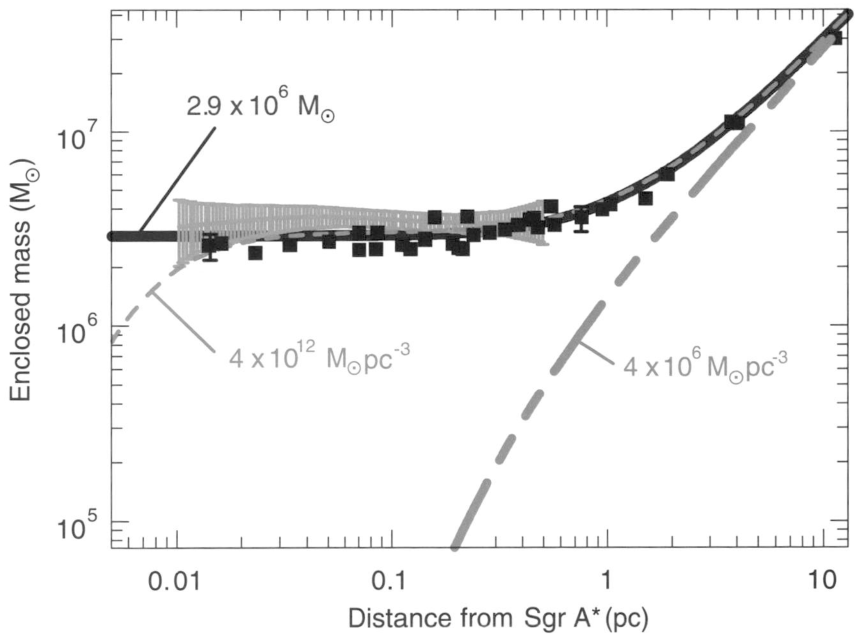

FIGURE 2.14 Evidence for a massive black hole at the galactic center, denoted Sgr A*. The data points are estimates of the distribution of mass, as determined from the motions of stars close to Sgr A*. The data (filled blue rectangles and light blue bars) are consistent with a 2.9 million solar-mass black hole (thick red curve) or a hypothetical very dense dark cluster (thin dashed green curve) that can be ruled out on theoretical grounds. The data rule out that the observed motions are caused by the gravitational field of the observed stellar cluster (short and long dashed green curve). Courtesy of R. Genzel (Max-Planck-Institut für extraterrestrische Physik).

FIGURE 2.15 Some of the data that provide strong evidence for the presence of a supermassive black hole in the center of the nearby spiral galaxy NCG 4258. The top panel is the actual image of the point-like maser clouds constructed from very long baseline interferometry (VLBI) data having a resolution of 200 microarcsec (with a wire grid depicting unseen parts of the disk). Also shown is the image of the continuum emission at 1.3-cm wavelength caused by synchrotron radiation from relativistic electrons emanating from the position of the dynamical center (black dot). The central mass required to gravitationally bind the system is 39 million solar masses. Since all the mass must be within the inner boundary of the molecular disk of about 0.13 pc, this mass is probably in the form of a supermassive black hole. The bottom panel shows on a larger scale the synchrotron emission that arises from relativistic electrons ejected along the spin axis of the black hole. Courtesy of J. Moran and L. Greenhill (Harvard-Smithsonian Center for Astrophysics), and J. Herrnstein (National Radio Astronomy Observatory).

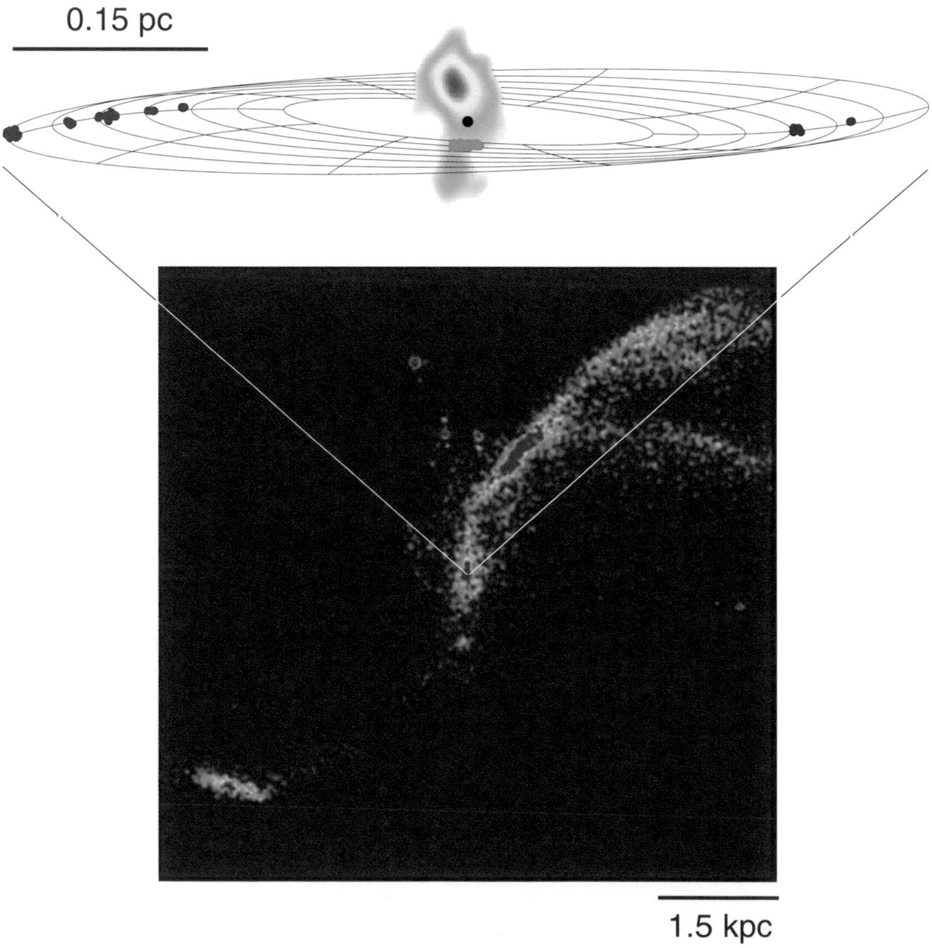

Observations with HST have confirmed that most nearby galaxies harbor supermassive black holes in their nuclei.

How do these supermassive black holes form and evolve? Do they grow from stellar "seeds" or do they originate at the very beginning of the formation of a galaxy? These key questions are ripe for a frontal attack now. Addressing them will require the observation of active galactic nuclei (AGNs) when they first turn on, over the entire electromagnetic spectrum. With its enormous sensitivity in the infrared, NGST will be able to detect AGNs out to redshifts beyond 10. Radiation emitted in the thermal infrared will be redshifted into the band detectable by SAFIR. The EVLA will detect much longer wavelength radio emission from AGNs to redshifts beyond 5. Constellation-X will be able to observe the first quasars even if they are heavily obscured by dust. EXIST will make a census of obscured, low-redshift AGNs over the whole sky; this sample can be compared with younger AGNs, seen at high redshifts by Constellation-X, to study how the AGNs evolve. In this case the energies of the most penetrating hard x rays will be conveniently shifted by the expansion of the universe into the energy region of maximum sensitivity of the telescope. Furthermore, by observing the spectrum of hot gas as it disappears into supermassive black holes, Constellation-X will provide a laboratory for studying the physical processes occurring near the event horizons of black holes under conditions that differ substantially from those near stellar-mass black holes.

In a tremendously scaled-up version of the process of mass ejection from disks around protostars, massive black holes not only accrete material but also eject from their vicinity powerful jets at nearly the speed of light (Figure 2.16). This highly relativistic material is thought to generate extremely energetic photons, with frequencies more than 100 billion times that of visible light. VERITAS has the power to detect individual photons of this radiation interacting with Earth's atmosphere, and can therefore probe the relativistic particle acceleration occurring near these massive black holes. Observing somewhat less energetic photons, GLAST will help determine how jets are powered and confined. ARISE has the spatial resolution to resolve the base of the jet and thereby provide a complementary probe of the acceleration region.

Galaxy mergers are inferred to be common, and it is quite possible that the massive black holes in their nuclei would merge as well. Such a cataclysmic event would produce powerful gravity waves that could be detected by LISA out to very large distances (redshifts up to at least 20). This gravitational radiation would be detectable for up to a year before the actual merger, enabling accurate prediction of the final event so that it could be observed by telescopes sensitive to the entire range of electromagnetic radiation. Observation of such a merger would provide a unique test of Einstein's theory of general relativity in the case of strong gravitational fields. Further discussion of what scientists can learn about black holes can be found in the physics survey report *Gravitational Physics: Exploring the Structure of Space and Time* (NRC, 1999).

Galactic nuclei can become extremely luminous as a result of intense bursts of star formation or the presence of a supermassive black hole.

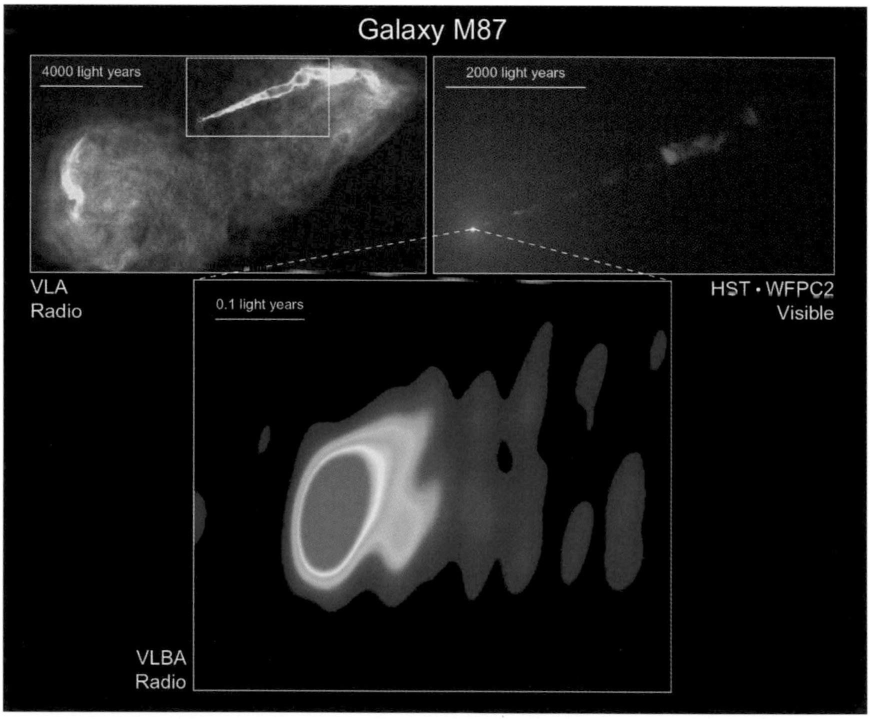

FIGURE 2.16 The jet produced by the central black hole in the galaxy M87. The Very Large Array (VLA) image at the upper left shows the radio emission powered by the jet. The Hubble Space Telescope (HST) image at the upper right shows the narrow jet at similar resolution. Finally, the Very Long Baseline Array (VLBA) image at the bottom, with more than 100 times the resolution of the HST image, is the closest view of the origin of such a jet yet obtained. Courtesy of NRAO, STScI, W. Junor (University of New Mexico), J.A. Biretta and M. Livio (STScI), and NASA. Reprinted by permission from *Nature* 401:891-892, copyright 1999 Macmillan Magazines Ltd.

These "starbursts" may be associated with the initial formation of the galaxy, or they may be triggered by an interaction with another galaxy. Starbursts are of great interest because they represent an extreme form of star formation that is not understood; for example, it is not known whether they produce the same distribution of stellar masses as that observed in our galaxy. Distinguishing starbursts from supermassive black holes is complicated by the fact that AGNs are often shrouded in dust, so that much of the direct emission is hidden from view. Long wavelengths penetrate the dust more readily, so the EVLA, SAFIR, and NGST with an extension into the thermal infrared are all suitable for separating the two phenomena. Very-high-energy photons can also penetrate the dust, so Constellation-X and EXIST will provide relevant data as well.

Active galactic nuclei may be the source of ultrahigh-energy cosmic rays (gamma-ray bursts and intergalactic shocks have also been suggested as the source of these enigmatic particles). These cosmic rays are generally assumed to be protons that have been accelerated to very high energies. The energies are so large—equivalent to the energy of 1 billion to 100 billion protons at rest—that these cosmic rays can propagate only a limited distance before losing their energy through interactions with the cosmic microwave background radiation. Ongoing experiments with the Fly's Eye in Utah and proposed experiments with the Southern Hemisphere Pierre Auger Observatory project will add greatly to our knowledge of these cosmic rays, particularly if the experiments are able to identify their sources.

THE UNIVERSE

Observations by NGST should witness the first light from distant galaxies. Long before the stars that emitted this light were formed, the matter making up the galaxies had to accumulate from the intergalactic medium. This process of galaxy formation occurred within the background of an expanding universe. How has the universe evolved through cosmic time? How did structures such as galaxies and clusters of galaxies develop in the expanding universe? Finally, observations show that not all the matter that makes up galaxies and clusters of galaxies is visible: What in fact is the composition of the universe?

THE EVOLUTION OF THE UNIVERSE

Evidence indicates that somewhat more than 10 billion years ago the universe was created in a titanic explosion—the Big Bang. What may have preceded this event is unknown. The Big Bang theory allows us to trace the evolution of the universe back to a time when it was just a soup of elementary particles—a few microseconds after the beginning. Researchers have promising ideas that would enable extending understanding back to a time before particles existed, when even the largest objects in the universe were quantum fluctuations. How has the universe expanded since the Big Bang? Astronomers measure the expansion of the universe through the redshift of the radiation observed. The greater the redshift of light from an observed object, the more the universe has expanded since that radiation was emitted. The relationship between the redshift and time—the calibration of the cosmic clock—determines how long ago the radiation was emitted (see Figure 2.12). Using the speed of light to convert time to distance, this relationship can be also be used to determine the geometry of the universe (whether space is flat or curved). The current time scale for the expansion is set by a parameter known as the Hubble constant, which gives the relation between redshift and distance. Using HST and other telescopes, it has been possible to establish the value of the Hubble constant with an accuracy approaching 10 percent.

In order to derive the age of the universe from the measured value of the Hubble constant, it is necessary to know how the expansion has accelerated or decelerated with time. The history of the expansion of the universe depends on the total density of matter in the universe (both ordinary matter and dark matter) and on the possibly non-zero "cosmological constant," which might characterize a sort of "dark energy" in the universe. These parameters determine the geometry of the universe and its ultimate fate, whether it will expand forever or eventually recollapse. Theory suggests that the geometry of the universe is flat; in this case, the total density of matter and energy is said to have its "critical" value. Observations of distant clusters of galaxies indicate that the density of matter is about 30 percent of the critical value.

One of the most exciting developments of the past decade has been the discovery that the cosmological constant may not be zero—our universe appears to be filled with dark energy. This discovery is based on two independent sets of observations. First, astronomers have found a way to determine the luminosity of Type Ia supernovae from the rate at which their light declines. Knowledge of the luminosity enables the determination (or calculation) of the distance to such a supernova by measuring its brightness. The results show that distant supernovae appear fainter than expected, suggesting that the expansion of the universe is accelerating. When combined with other data, the observations of supernovae lead to the conclusion that dark energy makes up perhaps 70 percent of the total density of matter and energy. Second, observations of fluctuations in the cosmic microwave background (discussed below) strongly suggest that the universe is indeed flat, so that the total density of matter and energy is at the critical value. Since estimates of the masses of clusters of galaxies show that the matter density of the universe has only about 30 percent of the critical value, it follows that the dark energy must make up the remaining 70 percent. Together with the value of the Hubble constant determined above, the estimated values of the matter and energy densities yield an age for the universe of about 14 billion years.

During this decade, observers and theorists will work to understand and extend these observations. Confirmation that dark energy exists, with a density that rivals that of matter, would be a physical discovery of the most fundamental significance. Planned observations of the cosmic microwave background will provide more accurate values of the cosmological parameters, including the density of ordinary matter. This value of the matter density, when compared to an equally precise determination derived from a measurement of the primeval deuterium abundance, will allow a fundamental consistency test of the standard cosmology. Recent measurements of the deuterium abundance in distant galaxies indicate that this test is feasible; however, a definitive measurement of deuterium is still needed. NGST will permit the observation of many supernovae at high redshifts, to confirm whether the universe is actually accelerating. Discovery of a much larger number of supernovae with LSST, followed up by more sensitive and precise measurements from ground- or space-based telescopes, will permit the cosmic clock to be calibrated with much greater precision. It should then be possible to determine whether the cosmological constant is really constant, as Einstein assumed, or evolving with time, as some current theories suggest.

THE EVOLUTION OF STRUCTURE IN THE UNIVERSE

The seeds of the structure of the universe down to the scale of galaxies, and probably even smaller, were planted by tiny quantum fluctuations in the first instants of the Big Bang. In order to study how the large-scale structure in the universe grew from these seeds, it is necessary to study how galaxies are distributed in space today. Surveys of galaxies carried out more than a decade ago revealed large voids where few galaxies were visible, and other regions where the density of galaxies was enhanced on scales up to 300 million light-years in extent. Surveys of galaxies during the past decade have shown that this appears to be the limiting scale on which large fluctuations in density occur: On larger scales, the universe appears to be smooth. Surveys under way now, particularly the Sloan Digital Sky Survey, will provide a far more accurate map of the distribution of galaxies in the nearby universe.

Direct evidence for the early fluctuations that led to this structure is imprinted on the oldest radiation in the universe, the cosmic microwave background (CMB). This radiation was emitted at a redshift of about 1,000, or a time only several hundred thousand years after the Big Bang, when the temperature of the radiation was somewhat less than that at the surface of the Sun. Today, the temperature of the background radiation is 1,000 times lower, just 3 degrees above absolute zero, having been cooled by the expansion of the universe. This radiation was observed with remarkable accuracy by the Cosmic Background Explorer (COBE), launched in 1989. Data from this satellite showed that the radiation had the theoretically predicted spectrum of a blackbody. COBE data also revealed tiny spatial ripples in the intensity of the radiation (Figure 2.17), indicative of density fluctuations that could lead to the observed large-scale structure of the universe. This set of satellite observations provided, for the first time, direct experimental evidence for a basic paradigm of scientists' cosmological speculations and established the quantitative basis for all subsequent work in this field.

By design, the COBE satellite had very low angular resolution, and therefore it was able to measure structure in the background radiation only on the largest scales. The characteristics of the background radiation on smaller scales depend on the matter and energy content of the universe; in concert with studies at lower redshifts, such as the Sloan Digital Sky Survey and searches for supernovae, these data can be used to determine all the fundamental properties of the universe, including its age and the amount of matter and energy it contains. Recent observations imply that the total density of matter and energy is very close to what is needed to make the geometry of the universe flat (see Figure 2.18). NASA's MAP, the European Space Agency's Planck Surveyor satellite, the ground-based Cosmic Background Imager, and future balloon observations will dramatically increase the sensitivity of studies of the background radiation. In addition to measuring the fundamental cosmological parameters with great precision, these missions will provide stringent tests of current cosmological theories. Ground-based studies will measure the distortion of the spectrum of the background radiation caused by the hot gas in intervening clusters of galaxies. Combined with observations by Constellation-X of the properties of this hot gas, these observations will enable researchers to determine the distances to these clusters, constrain the value of the Hubble constant, and probe the large-scale geometry of the universe.

One aspect of the cosmic microwave background that these missions will only begin to investigate is its polarization. Gravitational waves excited during the first instants after the Big Bang should have produced effects that polarized the background radiation. More precise measure-

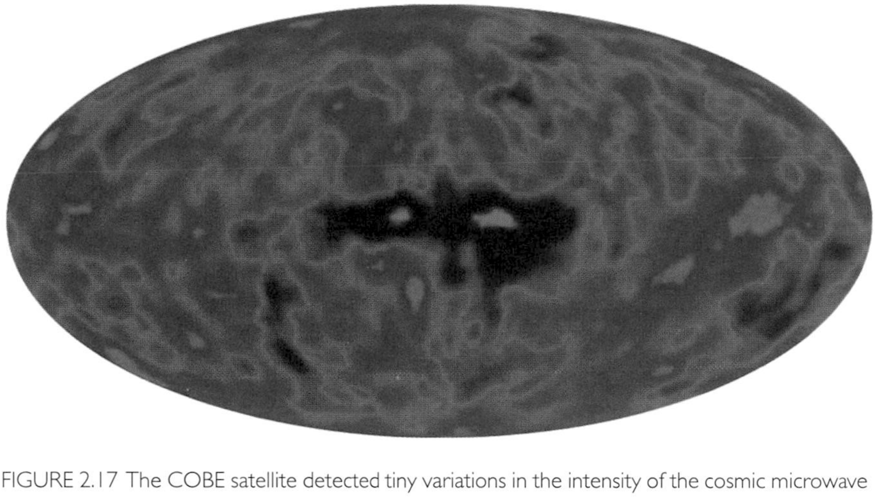

FIGURE 2.17 The COBE satellite detected tiny variations in the intensity of the cosmic microwave background. The amplitude of the temperature fluctuations is only about 0.00001 K, which reflects the smoothness of the universe at the time this radiation was emitted, and dramatically confirms the theoretical expectation that the universe began from a dense, hot, highly uniform state. The COBE data sets were developed by NASA's Goddard Space Flight Center under the guidance of the COBE Science Working Group and were provided by the National Space Science Data Center.

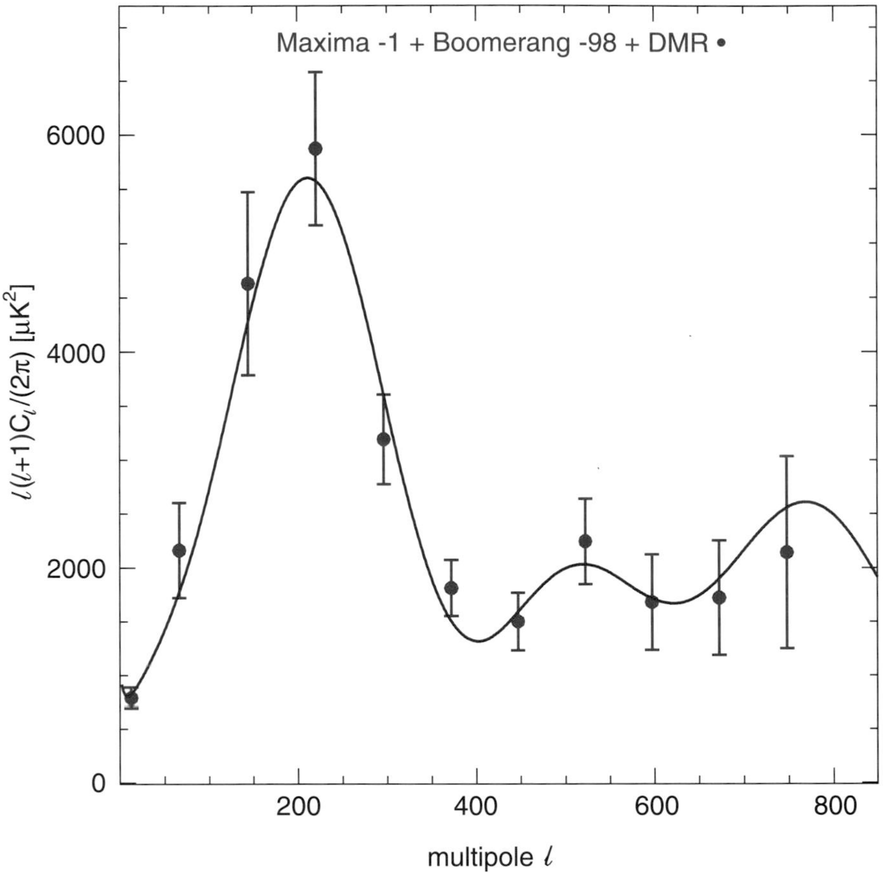

FIGURE 2.18 The spectrum of the primordial sound produced by the Big Bang. The sound waves can be observed through the fluctuations they produce in the temperature of the cosmic microwave background. Plotted is the mean-square temperature difference between two points in the sky as a function of their angular separation parameterized by the multipole number ℓ (angular separation ~ 180 degrees/ℓ). The observations were made with the BOOMERANG and MAXIMA balloon-borne telescopes; data from the COBE differential microwave radiometers (DMR) are also included. The peak in the spectrum at about 1 degree ($\ell \sim 200$) indicates that the universe is nearly spatially flat. The data can be fit well by models (such as that shown by the solid blue curve) in which only a small fraction of the matter is normal baryonic matter. Courtesy of the BOOMERANG and MAXIMA collaborations.

ments of the properties of this polarization—to be made by the generation of CMB missions beyond Planck—will enable a direct test of the current paradigm of inflationary cosmology, and at the same time they will shed light on the physics of processes that occurred in the early universe at energies far above those accessible to Earth-bound accelerators.

COMPOSITION OF THE UNIVERSE

Ordinary matter is made up of the same atoms as are known to us on Earth. The nucleus of an atom consists of protons and neutrons. The electrons encircling the nucleus are equal in number to the protons, although some of these electrons are stripped from the atom if the atom is ionized. Atoms can combine together into molecules, which in turn combine together to form all the matter we see on Earth. Atoms can produce light, and by observing light from stars astronomers have concluded that the stars, too, are made up of atoms. But when astronomers observe larger objects, such as the outer parts of galaxies or entire clusters of galaxies, they have found that the amount of matter they see in glowing gas and stars is not enough to hold these objects together by gravity. They therefore have postulated a form of matter too faint to see through its radiation: dark matter.

The current state of knowledge of the composition of the universe is shown in Figure 2.19. As discussed above, recent observations have suggested that the total density of matter and energy is the critical value necessary for a flat universe. Of this total critical value, about two-thirds is dark energy, whose nature is unknown, and one-third is matter. Ordinary matter is about 5 percent of the total, and luminous stars make up only about 0.5 percent. Where is the ordinary matter that is not in luminous stars? A leading contender for at least some of this missing ordinary matter is hot intergalactic gas, and Constellation-X will test this hypothesis. An even greater mystery is the nature of the matter that is not made up of atoms—the dark matter. Some of this matter is composed of neutrinos left over from the Big Bang. Although the uncertainty in their mass makes it difficult to determine exactly how much, astrophysical observations suggest that neutrinos do not account for the bulk of the dark matter. The rest is believed to be in the form of dark matter particles or objects that move relatively slowly, and are therefore called "cold" dark matter. Determination of the nature of this cold dark matter is one of the great unsolved problems in modern astrophysics.

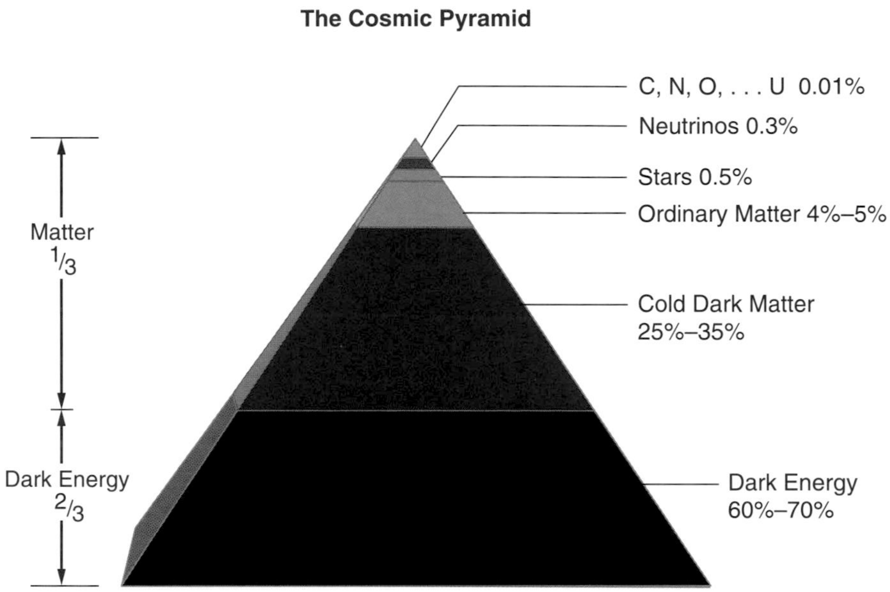

FIGURE 2.19 The makeup of our universe. Two-thirds of the matter and energy in the universe is in the form of a mysterious form of dark energy that is causing the expansion of the universe to speed up, rather than slow down. The other third is in the form of matter, the bulk of which is dark and which scientists believe is composed of slowly moving elementary particles (cold dark matter) remaining from the earliest moments after the birth of the universe. All forms of ordinary matter account for only about 5 percent of the total, of which only about one-tenth is in stars and a very tiny amount is in the periodic table's heavier elements (carbon, nitrogen, oxygen, and so on). The idea of particle dark matter was reinforced by recent indications that neutrinos have mass and thereby account for almost as much mass as do stars. Adapted from a drawing courtesy of M. Turner (University of Chicago).

The large-scale distribution of the dark matter can be studied through observations of gravitational lensing. Studies of gravitational lensing have given astronomers their best look at the distribution of dark matter both in clusters of galaxies and around some individual galaxies. In this decade, surveys of galaxies over vast areas of the sky with LSST and other telescopes will provide lensing data that describe the dark matter distribution over supercluster scales—information crucial for understanding the growth of large-scale structure.

Two leading possibilities for the makeup of dark matter are (1) elementary particles left over from the earliest moments of creation and (2) objects of stellar mass (massive compact halo objects, or MACHOs). It is a mark of the uncertainty in this field that these two candidates differ in mass by more than 57 orders of magnitude.

Theorists predicted that MACHOs, though too faint to be detected by their own emission, could be detected by gravitational lensing as well: The light of the background star would be amplified as the MACHO passed in front of the star. During the past decade, several groups independently detected this phenomenon, which is called microlensing because the mass of the lens is so small compared with that of galaxies (Figure 2.20). The nature of the MACHOs is a significant mystery: Are they stars made up of ordinary matter, or are they objects made up of an exotic form of matter? Accurate determination of their masses would help resolve this question, but to date, definitive measurements have not been possible; the best estimate is that the typical mass of a MACHO is somewhat less than a solar mass. By resolving the apparent motion of the stars that are imaged by the MACHOs, SIM will measure the masses of the MACHOs. Studies of microlensing have had several important spinoffs, including resolution of the surface of the star being lensed, and demonstration that it should be possible to detect planets as small as Earth through microlensing observations, as discussed in "The Formation and Evolution of Planets" section of this chapter.

As yet it is unclear how much MACHOs contribute to the dark matter in the Galaxy. If MACHOs are made of ordinary matter, then they cannot account for the bulk of the dark matter known to exist in the universe or even in our own galaxy. As a result, a number of efforts are under way in laboratories around the world to discover the particle dark matter that may be holding our own Milky Way together. There are two important

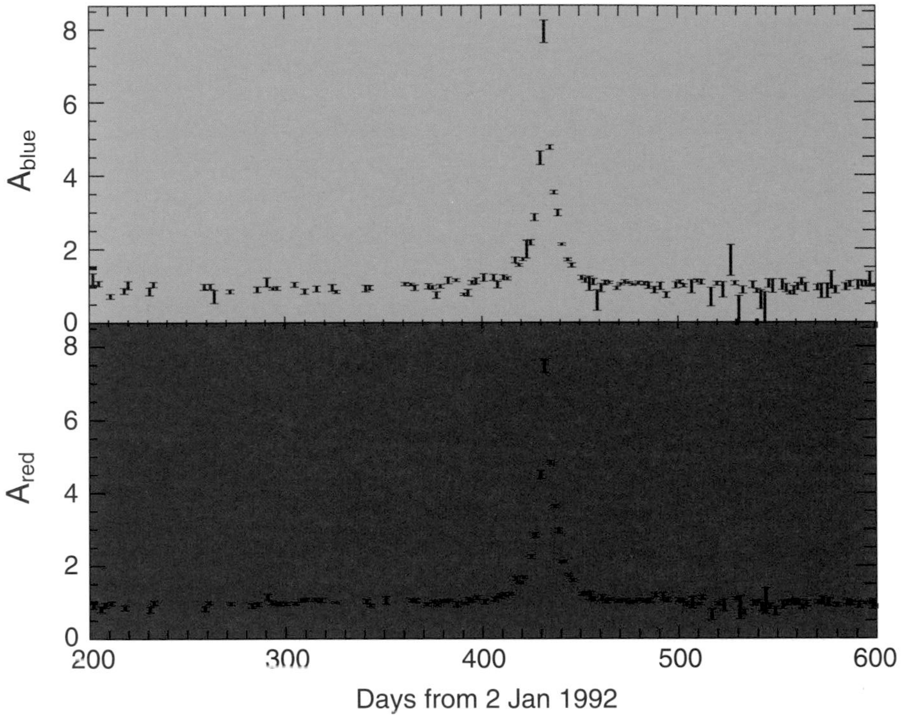

FIGURE 2.20 The first gravitational microlensing light curve, showing the amplification of the light of a background star by the gravitational field of an intervening object. These intervening objects, of unknown nature, may contribute to the dark matter in the Galaxy. The similarity of the curves in red light and blue light helps confirm that the brightening is caused by gravitational lensing. Courtesy of the MACHO collaboration. Reprinted by permission from *Nature* 365:621-623, copyright 1993 Macmillan Magazines Ltd.

ongoing efforts in the United States: (1) the Cryogenic Dark Matter Search II, a search for a particle with roughly atomic mass called the neutralino, and (2) the U.S. Axion Experiment, a search for an extremely light dark matter particle called the axion. The existence of the neutralino is a prediction of superstring theory, a bold and promising attempt to unify gravity with the other forces of nature. The discovery that neutralinos or axions are the dark matter that binds our own galaxy would shed light not only on the astrophysical dark matter problem, but also on the unification of the fundamental forces and particles of nature.

3

The New Initiatives: Building on the Current Program

INTRODUCTION

Astronomers seek to understand phenomena as diverse as the formation of planets, the origin of solar activity, the evolution of black holes, and the large-scale structure of the universe. Electromagnetic radiation—radio, infrared, optical, ultraviolet, x-ray, and gamma-ray—provides most of the information we have on these distant phenomena, but other messengers can contribute as well: Energetic charged particles (the cosmic rays) carry information about where they were accelerated, neutrinos tell us about the deep interior of stars, and gravitational waves may reveal how some black holes form. This chapter describes how the recommended new initiatives grew out of the existing program, how they complement each other, and how they will address the major scientific questions in astronomy. The characteristics of the major and moderate new initiatives are listed in Tables 3.1 and 3.2. More detailed descriptions of the facilities and the science they will accomplish can be found in *Astronomy and Astrophysics in the New Millennium: Panel Reports* (NRC, 2001).

The committee emphasizes that telescopes alone do not lead to a greater understanding of the universe. So that maximum benefit can be obtained from the current and proposed new facilities, the committee recommends a vigorous and balanced program of astrophysical theory, data archiving and mining, and laboratory astrophysics. One of the key recommendations is the establishment of theory challenges to be associated with most new major and moderate programs. These challenges should describe theoretical problems that are ripe for progress, relevant to the planning and design of the program, and essential to the interpretation and understanding of its results in the broadest context. The specific theory challenges tied to each mission and project should be determined by the informed astronomical community—probably through ad hoc panels, drawn from the theory community and convened for this purpose. However, to illustrate the concept the committee gives examples of possible theory challenges below in the discussion of the new initiatives.

THE ULTRAVIOLET, OPTICAL, AND INFRARED WINDOWS ONTO THE UNIVERSE

Ultraviolet, optical, and infrared (UVOIR) observations provide extremely important sources of information about the universe. Stars,

TABLE 3.1 Characteristics of Prioritized Major Initiatives

Project	Wavelength or Energy Coverage	Increase in Sensitivity[a]	Spatial Resolution (arcsec)[b]	Spectral Resolving Power[c]
NGST	0.6 to 27 μm (goal)	100 to 600	$0.05\left(\frac{\lambda}{2\ \mu m}\right)$	5 to 5,000
GSMT	0.3 to 25 μm	10	$0.008\left(\frac{\lambda}{1\ \mu m}\right)$	3 to 10^5
Con-X	0.25 to 40 keV	100	15	300 to 5,000
EVLA	3 mm to 100 cm	10	$0.003\left(\frac{\lambda}{3\ mm}\right)$	10 to 10^7
LSST	0.3 to 1 μm 0.3 to 2.4 μm (goal)	20 to 60	0.6	3 to 100
TPF	3 to 30 μm	n.a.[d]	$0.00075\left(\frac{\lambda}{3\ \mu m}\right)$	3 to 300
SAFIR	30 to 300 μm	100 to 300	$0.8\left(\frac{\lambda}{30\ \mu m}\right)$	5 to 10^3

[a]Increase in sensitivity compared with existing or other planned facilities.

[b]For entries with a (λ/a) term included, the spatial resolution is for wavelength $\lambda > a$. The number in front of the parenthesis therefore corresponds to the best spatial resolution.

[c]Spectral resolving power is defined as $\lambda/\Delta\lambda$.

[d]n.a., not applicable, is assigned to TPF because there is no existing facility with which to compare the sensitivity. This unique facility will work at spatial resolutions not currently accessible in this wavelength range.

and the interstellar gas and dust that make stars, emit most of their radiation at these wavelengths. The atoms and ions in the interstellar medium and in stellar atmospheres have rich UVOIR spectra that can be used to probe the density, dynamics, magnetic field structure, temperature, and elemental abundances of astronomical objects. Infrared radiation penetrates the obscuring dust that surrounds regions of star and

TABLE 3.2 Characteristics of Prioritized Moderate Initiatives

Project	Wavelength or Energy Coverage	Increase in Sensitivity[a]	Spatial Resolution (arcsec)[b]	Spectral Resolving Power[c]
GLAST	10 MeV to 300 GeV	>40	600	50
LISA	10^{-4} to 10^{-1} Hz[d]	n.a.[e]	1,800	3×10^4
AST	0.3 to 35 μm	300	$0.05 \left(\frac{\lambda}{1\ \mu m}\right)$	10^3 to 10^6
SDO	0.02 to 1 μm	25	1	100 to 6×10^4
CARMA	860 μm to 1 cm	4 to 12	$0.1 \left(\frac{\lambda}{1\ mm}\right)$	3 to 10^7
EXIST	5 to 600 keV	1,000	300	100
VERITAS	100 to 10^4 GeV	10	180	10
ARISE	3 mm to 4 cm	n.a.[e]	$10^{-5} \left(\frac{\lambda}{3\ mm}\right)$	20 to 10^7
FASR	1 to 100 cm	n.a.[e]	$0.67 \left(\frac{\lambda}{1\ cm}\right)$	30 to 100
SPST	200 μm to 1 mm	10	$3 \left(\frac{\lambda}{200\ \mu m}\right)$	10 to 10^7

[a]Increase in sensitivity compared with existing or other planned facilities. For EXIST the comparison is with a previous survey by HEAO-1.

[b]For entries with a (λ/a) term included, the spatial resolution is for wavelength $\lambda > a$. The number in front of the parenthesis therefore corresponds to the best spatial resolution.

[c]Spectral resolving power is defined as $\lambda/\Delta\lambda$.

[d]Frequency of gravitational waves. All other recommended projects detect electromagnetic waves.

[e]n.a., not applicable, is assigned to LISA, ARISE, and FASR because there are no existing facilities with which to compare the sensitivity. LISA will be a unique facility operating at frequencies (see footnote above) not accessible to the LIGO experiment. ARISE will operate at unprecedented spatial resolutions. FASR will observe the Sun over an unprecedented frequency range and at spatial and temporal resolutions beyond current capabilities.

planet formation and many galactic nuclei. Although stars generally radiate mostly at optical and ultraviolet wavelengths, the expansion of the universe transforms such radiation from very distant galaxies to infrared wavelengths. Consequently, a number of the new initiatives cover the infrared portion of the spectrum.

Advances in UVOIR technology offer the promise of enabling great leaps in our understanding of the universe. Typically, these advances bring greater sensitivity—the ability to detect faint signals—and greater angular resolution, the ability to see fine detail in distant objects. Increased sensitivity requires more collecting area, which is most easily achieved with large filled-aperture telescopes. The angular resolution also improves with the diameter of the telescope mirror. However, to achieve the extremely high angular resolution needed for some of the long-range scientific goals described in Chapter 2 would require excessively large mirrors, up to thousands of kilometers in size. For these angular resolutions filled-aperture telescopes are not practical, and interferometers are used instead. Interferometers combine the photons from two or more telescopes to produce the equivalent angular resolution of a single telescope whose diameter equals the maximum separation of the interferometer telescopes. The cost of using interferometers is reduced sensitivity and ambiguities in image reconstruction. The latter can be corrected by reobserving with many different baseline separations and orientations. Interferometers also can cancel the light from the central star in a planetary system so that astronomers can see the relatively dim planets nearby.

Both space- and ground-based telescopes are needed to open the UVOIR window onto the universe. Ground-based telescopes have the advantage that they are generally less expensive and much easier to maintain and upgrade. Space-based telescopes have the advantage that they are free from the distortion, absorption, and background emission of the atmosphere and allow observations in many UVOIR wavelength bands not accessible from the ground: for example, the far ultraviolet (0.1 to 0.3 μm), many regions in the infrared (1 to 30 μm), and the entire far infrared (30 to 300 μm). Even at wavelengths that do reach the ground, the turbulence in the atmosphere blurs the angular resolution of ground-based telescopes. Recent advances in adaptive optics have substantially reduced this problem in the infrared, but it persists in the optical band, particularly for wide-field imaging. Together, ground- and space-based telescopes enable a comprehensive attack on many of the fundamental questions in astronomy.

LARGE FILLED-APERTURE OPTICAL AND INFRARED TELESCOPES: NGST AND GSMT

At the present time, the premier UVOIR telescopes are the Hubble Space Telescope (HST) in space and the Keck 10-m telescopes on the ground. When they are completed, the 8-m Gemini telescopes and the European Very Large Telescope (VLT) will provide powerful additions to the available ground-based telescopes (see Table 3.3). The committee's top recommendations for this decade are to dramatically increase the capability of UVOIR observations with the Next Generation Space Telescope (NGST) in space and the Giant Segmented Mirror Telescope (GSMT) on the ground. Both of these telescopes have filled apertures and are very substantial improvements on HST and Keck, respectively. Their large mirrors must be segmented, and fast-responding actuators must bring the segments into nearly perfect alignment. The 1990s saw revolutionary advances in the development of lightweight mirrors and segmented structures that will be put to use in constructing these new telescopes.

NGST (Figure 3.1) consists of a passively cooled, segmented telescope that will deploy to its full diameter of about 8 m once it is in space. It will orbit the Sun roughly a million miles from Earth. At present, its planned wavelength range is 0.6 to 27 μm. NGST will be far more capable than its space predecessors HST and SIRTF and its airborne predecessor SOFIA. Figure 3.2 compares the sensitivity of NGST with that of other space facilities at low spectral resolving power, which is appropriate in searches for distant galaxies and faint stellar objects. Much of its increase in sensitivity compared with previous space telescopes comes from its large aperture, which not only gathers more photons from each source but also reduces the number of photons from the background by virtue of its greater angular resolution. Astronomical capability is defined in the 1991 survey, *The Decade of Discovery in Astronomy and Astrophysics* (NRC, 1991), in terms of the speed of an observation. Improvements in sensitivity and angular resolution make NGST roughly 1,000 times more capable than HST and SIRTF; its low temperature makes it up to a million times more capable than similar-size ground-based telescopes. The discovery potential of NGST is enormous. Having NGST's sensitivity extend to 27 μm would substantially improve its ability to study Kuiper Belt objects (KBOs) in the solar system, star formation and planet formation in our galaxy, and dust emission in galaxies out to a redshift of 3. Not only would this extension take full

TABLE 3.3 Existing and Approved Large Ground-Based Optical and Infrared Telescopes

Project	Commencement of Operations	Partners Involved	Percent U.S.	Apertures[a]
Gemini		United States, United Kingdom,		
Northern	1999	Canada, Chile, Australia,	52	2 × 8 m
Southern	2001	Argentina, Brazil	42	
Gran Telescopio Canarias	2003	Spain	0	1 × 10.4 m
HET	1998	United States,[b] Germany	90	1 × 9.2 m
Keck	1993, 1996	United States[b]	100	2 × 10 m
LBT	2002	United States,[b] Italy, Germany	50	2 × 8.4 m
Magellan	2000, 2002	United States,[b] Chile	90	2 × 6.5 m
MMT	1999	United States[b]	100	1 × 6.5 m
SALT	2002	South Africa, United States,[b] Poland, New Zealand, Germany, United Kingdom	20+	1 × 10 m
Subaru	1999	Japan	0	1 × 8 m
VLT	1999+	Europe	0	4 × 8 m

[a]Notation: 2 × 8 m denotes two telescopes with 8-m-diameter apertures.
[b]U.S. private or university telescopes.
SOURCE: Includes data from NRC (2000), Table 5.13.

advantage of the effort to cool the instrument, but NGST would also gain its greatest advantage over any ground-based telescope at the longer infrared wavelengths (see Figure 3.3).

Considerable progress has been made in developing the challenging technology required by NGST, including sensitive detectors, lightweight deployable primary mirrors, and control and image analysis systems. To enable NGST to reach its full potential, the committee recommends technology development to increase telemetry rates in spacecraft communication and for cryocoolers that enable detectors to operate at wavelengths longer than 5 μm.

GSMT complements NGST, both in technical capabilities and in its ability to probe distant galaxies and nearby star-forming regions. GSMT is a 30-m-class, ground-based, filled-aperture, segmented-mirror, optical and infrared telescope that will operate in the atmospheric windows over the wavelength range from 0.3 to 25 μm. Adaptive optics will give it diffraction-limited performance down to wavelengths as short as 1 μm. GSMT will complement NGST much as the Keck telescope has comple-

FIGURE 3.1 An artist's conception of the NGST mission. NGST, the highest priority for the new decade, will be a segmented, filled-aperture, 8-m-class telescope that will utilize a large sunshield and an orbit at least 1 million km from Earth to achieve very cold operating temperature. Its sensitive measurements, primarily in the infrared region of the spectrum, will revolutionize the fields of galaxy formation and evolution and star formation. Courtesy of J. Lawrence (NASA Goddard Space Flight Center).

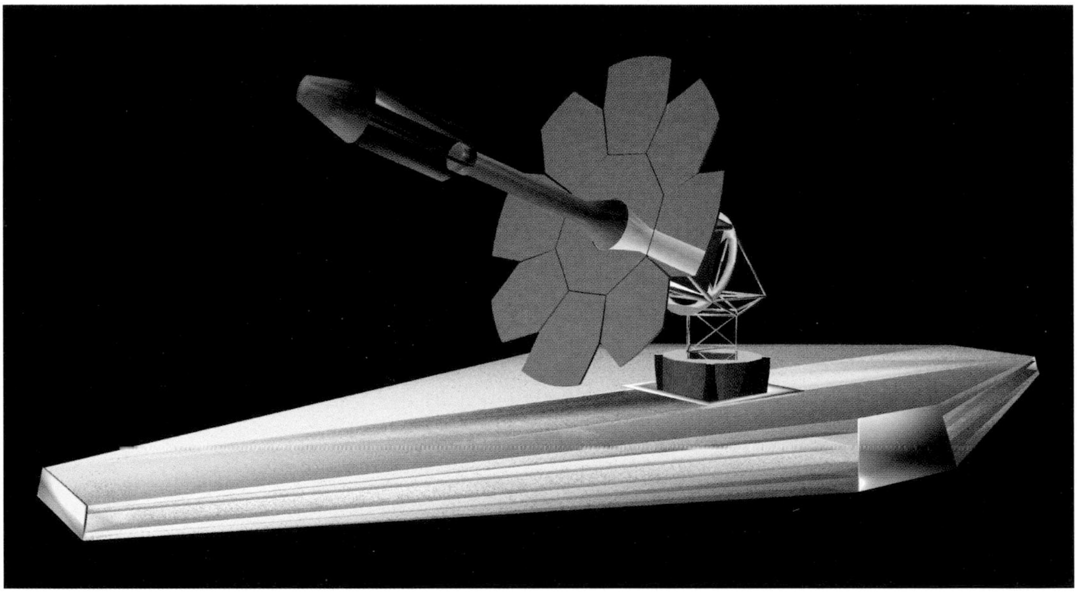

mented HST, by making studies with high spatial and spectral resolution of the sources seen by the smaller space telescope. Figure 3.3 shows a comparison of the sensitivity of GSMT and NGST at various spectral resolving powers, demonstrating the power of NGST at low spectral resolution and longer wavelengths and the power of GSMT at high spectral resolution and shorter wavelengths. Furthermore, with the ability to add new instrumentation, a step that is not possible with NGST, GSMT can evolve its capabilities and become increasingly powerful.

In agreement with the Panel on Optical and Infrared Astronomy from the Ground (see Chapter 2 of the *Panel Reports;* NRC, 2001), the committee believes that the 30-m scale of GSMT is the appropriate next step in the construction of large ground-based OIR telescopes. The more ambitious 100-m "OWL" telescope under development by the European

Southern Observatory represents an excellent opportunity for shared technology development and possibly for eventual U.S. collaboration. International participation in GSMT would offer strong benefits as well.

Experience with the Keck, Gemini, and Hobby-Eberly telescopes shows that the cost of large telescopes need not increase with a high power of the mirror diameter. Nevertheless, the large size of GSMT means that substantial advances in telescope design and adaptive optics will be required if it is to be built for a reasonable cost. The committee recommends that this work commence soon so that construction of the telescope can begin in this decade. Figure 3.4 shows the enormous gains in spatial resolution that are made possible by the use of adaptive

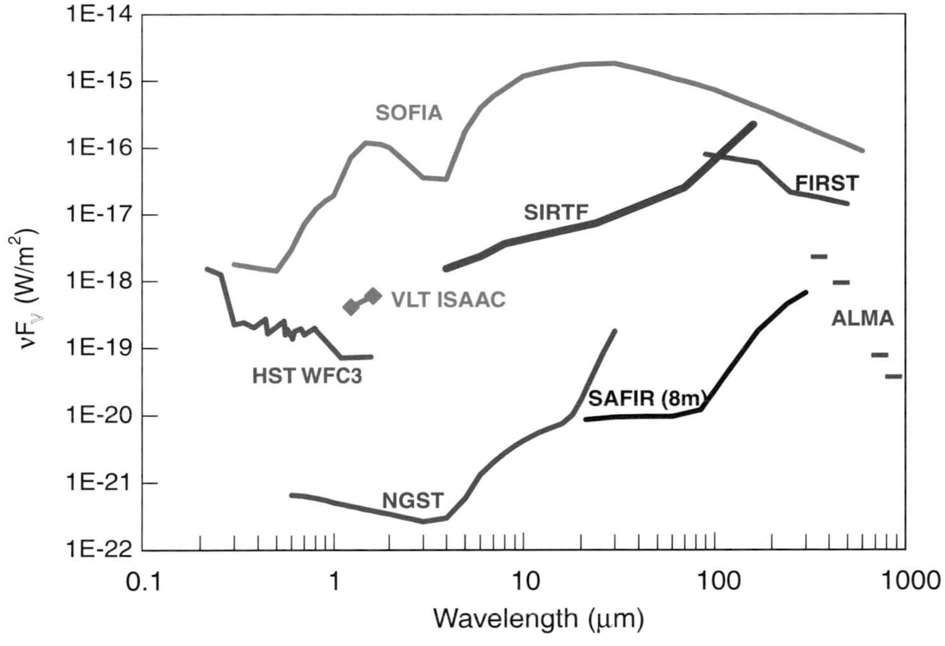

FIGURE 3.2 Relative performance of planned and recommended space initiatives, as well as the performance of the ground-based ALMA in the submillimeter wavelength band. The vertical axis denotes the 5σ flux detectable in 3 hours of integration; therefore better performance is *lower* on the figure. Estimated performance is based on the combination of photon and confusion noise appropriate to the high-galactic-latitude sky. Courtesy of G. Rieke (University of Arizona) and S. Beckwith (Space Telescope Science Institute).

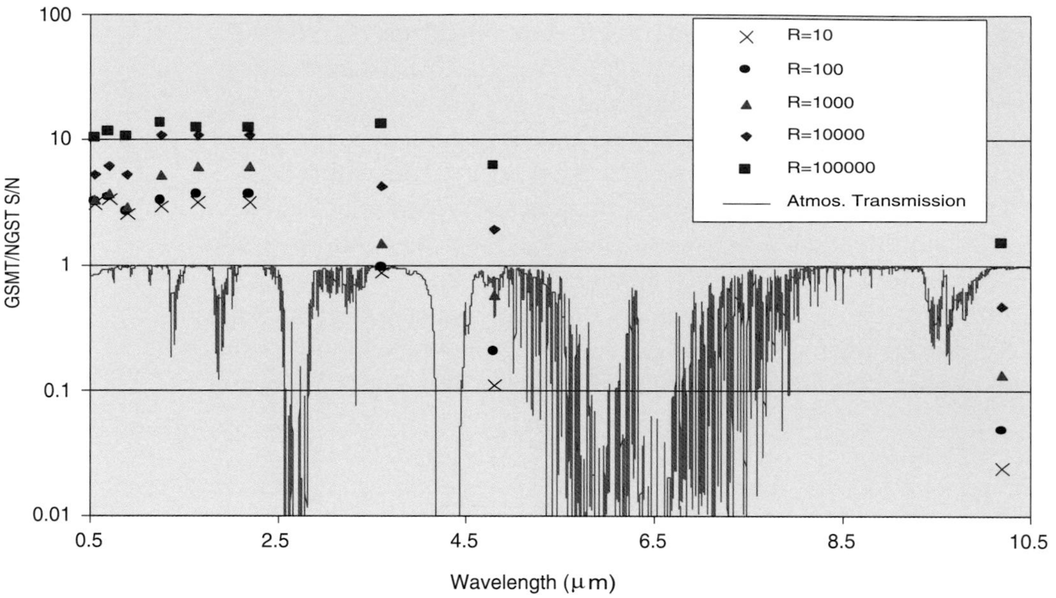

FIGURE 3.3 A comparison of GSMT to NGST for spectral resolving powers of R = 10 through 100,000 over a range of wavelengths. Plotted is the signal-to-noise ratio achieved in an observation of given duration on a given object (much fainter than the sky) for GSMT relative to that achieved with NGST. The comparison shows that GSMT is substantially more effective in obtaining high-resolution spectra of faint objects for short-wavelength radiation that penetrates the atmosphere. NGST is substantially more effective at longer wavelengths, at wavelengths blocked by the atmosphere, and for observations done at low spectral resolution. Courtesy of L. Ramsey (Pennsylvania State University).

optics on ground-based telescopes such as GSMT. GSMT requires a large investment of resources and offers an opportunity for partnership between national and university/independent observatories in producing and operating a world-class facility within the coordinated system of these two essential components of U.S. ground-based astronomy.

Together, NGST and GSMT will trace the formation and evolution of galaxies from the end of the "dark ages," when the first stars formed, until the present. While NGST's infrared capability will enable it to study that early epoch of the universe when clouds of hydrogen gas collapsed to form the first galaxies and stars, GSMT will be especially powerful in studying galaxies and intergalactic gas at a somewhat later period of cosmic history when most of today's stars and chemical elements were formed. NGST will observe the development of the clustering and

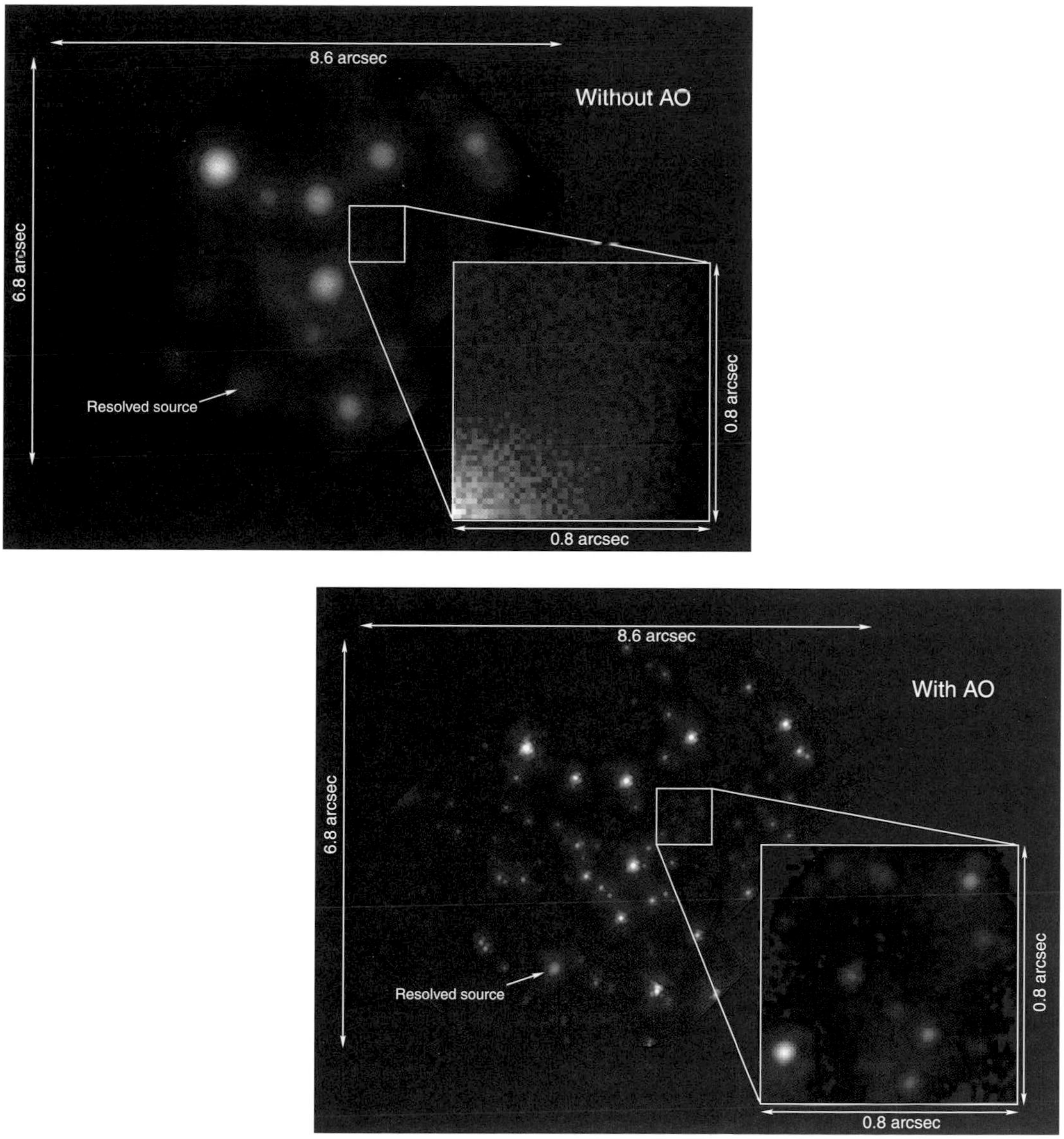

FIGURE 3.1 The power of adaptive optics (AO) as shown by comparison of a high-quality 2.2-µm ground-based image of the galactic center without AO and with AO. Adaptive optics corrects for the distortions caused by the turbulence in the atmosphere and results in an image of much higher resolution (diffraction-limited). Courtesy of the W.M. Keck Observatory Adaptive Optics Team. (This figure originally appeared in *Publications of the Astronomical Society of the Pacific* [Wizinowich, P., et al., 2000, vol. 112, pp. 315-319], copyright 2000, Astronomical Society of the Pacific; reproduced with permission of the Editors.)

merging of the first galaxies, together with the history of star formation in galaxies. The capabilities of GSMT to study relatively faint sources with high spatial resolution and adequate spectral resolution will provide an essential complement to NGST in understanding how galaxies form and evolve.

GSMT will have the capability of studying the nearest planetary systems and star-forming regions of the Milky Way Galaxy, utilizing high spatial and spectral resolution to directly observe massive planets like Jupiter and to study the innermost regions of protoplanetary disks to ascertain how stars grow and what causes the powerful outflows or winds from young protostars. NGST will be a powerful complement to GSMT in this endeavor by enabling study of the continuum emission from protoplanetary disks, particularly if its spectral imaging capability is extended to the thermal infrared.

These missions address major problems across a broad range of astrophysics. Possible theory challenges for NGST are

❖ *To develop an integrated theory of the formation and evolution of large-scale structure in the universe, Lyman-α clouds, galaxy clusters, and galaxies; and*

❖ *To understand supernovae—the mechanism of explosion, the spectra, and the light curves.*

A possible theory challenge for GSMT is

❖ *To develop models of star and planet formation, concentrating on the long-term dynamical co-evolution of disks, infalling interstellar material, and outflowing winds and jets.*

Each of these problems requires substantial efforts in numerical simulation as well as in basic theory. The simulations will be particularly valuable in making detailed comparisons of theoretical models with observations.

OPTICAL AND INFRARED SURVEYS: LSST

Telescopes like GSMT and NGST look at selected regions of the sky or study individual sources with high sensitivity. However, another type of telescope is needed to survey the entire sky relatively quickly, so that periodic maps can be constructed that will reveal not only the positions of target sources, but their time variability as well. The committee recommends the Large-aperture Synoptic Survey Telescope (LSST), a 6.5-m-class very-wide-field (~ 3 deg) telescope that will produce a deep (~ 24th magnitude in a single optical band) digital map of the visible sky every week. Not only will LSST carry out an optical survey of the sky far deeper than any previous survey (see Table 3.1), but it will also add the new dimension of time and thereby open up a new realm for discovery. By surveying the sky each week over a decade, LSST would revolutionize our knowledge of astronomical sources whose light varies on time scales of days to years. Such an experiment could locate 90 percent of all near-Earth objects down to 300 m in size, enable computation of their orbits, and permit assessment of their threat to Earth. It would discover and track objects in the Kuiper Belt, a largely unexplored, primordial component of our solar system. It would discover and monitor a wide variety of variable objects, such as the optical afterglows of gamma-ray bursts. In addition, it would find approximately 100,000 supernovae per year. Analysis of the data on these supernovae would shed light on the distribution of dark matter by determining the peculiar motions of galaxies, and it would provide valuable data on the evolution of stellar populations in galaxies with a wide range of ages. The faintest and most plentiful of this sample of supernovae could be followed up with other ground- or space-based telescopes to study the dynamics of the universe over the last half of its history. LSST would also provide valuable data on steady sources: By adding the data from different nights, it would be possible to develop maps of galaxies down to very faint magnitudes. Such maps would make it possible to infer the structure of dark matter on large scales from the way in which the dark matter distorts the images of the galaxies through weak gravitational lensing. A second-generation instrument with infrared detectors could generate a map of the sky that is 100 times deeper than that obtained with the Two Micron All Sky Survey (2MASS).

With its huge arrays of detectors, LSST will collect more than a trillion bits of data per day, and the rapid data reduction, classification, archiving, and distribution of these data will require considerable effort. The resulting database and data-mining tools will likely form the largest nonproprietary data set in the world and could provide a cornerstone for the National Virtual Observatory (see below). The construction and operation of LSST, together with the processing and distribution of the data, provide critical community service opportunities for an effective national organization for ground-based OIR astronomy.

Possible theory challenges for LSST are

❖ *To study the origin and fate of comets and asteroids, and their relation to the building blocks from which planets are formed; and*

❖ *To discover, interpret, and explain evidence of unexpected rare phenomena buried in the LSST database.*

THE TELESCOPE SYSTEM INSTRUMENTATION PROGRAM—TSIP

The committee's highest priority in the moderate-cost category of either space- or ground-based initiatives is the Telescope System Instrumentation Program (TSIP). Currently available resources for properly instrumenting the suite of 6- to 10-m-class ground-based telescopes (see Table 3.3) are woefully inadequate to fully exploit the potential of facilities available to U.S. astronomers. By substantially increasing the funding of facility instruments for the new generation of large-aperture telescopes at independent and university observatories, the NSF will encourage the continuation of substantial nonfederal investments, leverage their scientific productivity, and add new observing opportunities for the entire U.S. astronomical community. The philosophy of the TSIP is consistent with previous recommendations of *A Strategy for Ground-Based Optical and Infrared Astronomy* (the McCray report; NRC, 1995), which also recognized the importance of such a program. The TSIP applies to grants exceeding $1 million and does not replace the existing Advanced Technologies and Instrumentation or Major Research Instrumentation programs at the NSF.

- **The committee recommends that, in exchange for TSIP funds, private observatories provide an opportunity to the entire astronomy community to apply for telescope observing time whose value (based on amortized investment and operations) would amount to 50 percent of the granted funds.**

The justification for this recommendation is discussed in Chapter 2 of the *Panel Reports* (NRC, 2001). Through its administration of the TSIP, the NSF can assist national and private observatories in working together as a system so that they can achieve their full scientific potential.

The TSIP will provide the instrumentation and the telescope time to address a number of fundamental problems: for example, the structural and chemical evolution of galaxies out to redshifts of order 2, the history of star formation as a function of galaxy type and luminosity, the nature of gamma-ray bursts, the evolution of the stellar halo of our galaxy, and the physics of brown dwarfs. Bringing adaptive optics to these telescopes will enable astronomers to monitor changes in the atmospheres of Mars, the Jovian planets, and Titan; to study protoplanetary disks around newly formed stars; and to investigate the structure of active galactic nuclei near their central black holes.

FAR-INFRARED ASTRONOMY FROM SPACE: SAFIR

At wavelengths between 30 μm and 300 μm the atmosphere is so opaque that astronomical observations can be made only from airborne or space-based observatories. Table 3.4 includes the three space missions under construction or in development that cover this wavelength band: the Space Infrared Telescope Facility (SIRTF), the European Far Infrared Space Telescope (FIRST), and the airborne Stratospheric Observatory for Infrared Astronomy (SOFIA). SIRTF and SOFIA should be operational in 2002, whereas FIRST is scheduled to be launched in 2007. SIRTF will make pioneering observations of the infrared emission from young, distant galaxies and will bring tremendous advances in the understanding of brown dwarfs, ultraluminous infrared galaxies, and the dusty disks that surround stars. SOFIA will provide higher-spectral- and higher-spatial-resolution studies of bright SIRTF sources and will make major contributions to research on nearby regions of star formation and the interstellar medium of nearby galaxies.

TABLE 3.4 Existing and Planned Large Space-Based Ultraviolet, Optical, and Infrared Telescopes

Project	Scheduled Years of Operation	Nations Involved	Percent U.S.	Wavelength Band[a]
HST	1990 to 2010	United States, Europe	85	UV, O, IR
SIRTF	2002 to 2007	United States	100	IR, FIR
SOFIA	2002 to 2022	United States, Germany	80	IR, FIR, SMM
SIM	2006 to 2011	United States	100	O
FIRST	2007 to 2012	Europe, United States	10	FIR, SMM

[a]For the purposes of this table, ultraviolet (UV) = 0.1 to 0.3 μm, optical (O) = 0.3 to 1.0 μm, infrared (IR) = 1.0 to 30 μm, far-infrared (FIR) = 30 to 300 μm, and submillimeter (SMM) = 300 μm to 1 mm.

To take the next step in exploring this important part of the spectrum, the committee recommends the Single Aperture Far Infrared (SAFIR) Observatory, a passively cooled 8-m-class telescope that builds on the technology developed for NGST. As shown in Figure 3.2, SAFIR will be far more sensitive than FIRST, SOFIA, and SIRTF at these wavelengths and will provide 2 to 10 times the spatial resolution of these upcoming missions. The combination of its size, low temperature, and detector capability makes its astronomical capability about 100,000 times that of other missions and gives it tremendous potential to uncover new phenomena in the universe. SAFIR will complement ALMA, NGST, and TPF by providing sensitive coverage of the wavelengths that lie between the capabilities of these missions. A rational coordinated program for space optical and infrared astronomy would build on the experience gained with NGST to construct SAFIR, and then ultimately, in the decade 2010 to 2020, build on the SAFIR, TPF, and SIM experience to assemble a space-based, far-infrared interferometer.

SAFIR will study the birth and evolution of stars and planetary systems so young that they are invisible to optical and near-infrared telescopes such as NGST. Not only does the far-infrared radiation penetrate the obscuring dust clouds that surround these systems, but the protoplanetary disks also emit much of their radiation in the far infrared. Furthermore, the dust reprocesses much of the optical emission from the newly forming stars into this wavelength band. Similarly, the obscured central regions of galaxies, which harbor massive black holes and huge bursts of star formation, can be seen and analyzed in the far infrared. SAFIR will have the sensitivity to see the first dusty galaxies in the universe. For the studies of both star-forming regions in our galaxy and dusty galaxies at high redshifts, SAFIR will be essential in tying together information that NGST will obtain on these systems at shorter wavelengths and that ALMA will obtain at longer wavelengths.

A possible theory challenge for SAFIR is

❖ *To understand the origin and evolution of dust in the universe.*

INFRARED INTERFEROMETRY FROM SPACE: TPF

The past decade has seen enormous strides in infrared and optical interferometry. The Palomar Testbed Interferometer and the Infrared Spatial Interferometer are in operation now, and the Keck Interferometer, the Center for High Angular Resolution Astronomy Array, the Very

Large Telescope Interferometer (VLT-I), the Large Binocular Telescope (LBT), and the Space Interferometry Mission (SIM) are under development. SIM (see Table 3.4) will be the first space-based interferometer and will demonstrate in space the technique of "nulling" so essential in observing dim planets orbiting bright stars. A particular attraction of SIM is its dual capability: It enables both narrow-angle astrometry for detecting planets and wide-angle astrometry for mapping the structure of the Milky Way and other nearby galaxies. It is critical that an accuracy of a few microarcseconds for wide-angle measurements be achieved in order to address a wide variety of fundamental problems throughout this decade.

These projects lay the groundwork for a truly revolutionary and ambitious mission, the Terrestrial Planet Finder (TPF), which could start before the end of the decade if the precursor missions and technology development proceed successfully and in a timely fashion. As envisioned during the present study, TPF consists of four 3.5-m telescopes, flying in carefully controlled formations spaced tens of meters to 1 km apart, in an orbit far from the heating influence of Earth. Its solar shields will keep the telescope in shadow and will cool it to less than 40 degrees above absolute zero (–390 °F). This low temperature will give it far greater sensitivity than earthbound interferometers in its operating range of 3 to 30 μm. The greatest challenge facing TPF is the capability to enable study of the very faint radiation from an Earth-sized planet against the glare of the central star. Earth radiates roughly a million times less infrared radiation than does the Sun, and when viewed from the nearest star it is only about 1 second of arc in angle (roughly the angle made by a dime seen a mile away) away from the Sun. TPF must also be able to detect planets against the infrared background provided by circumstellar dust (exozodiacal light) in the planetary systems. Large ground-based interferometers such as Keck and LBT can resolve this emission, while SIRTF will survey all the potential TPF targets to levels approaching the strength of the zodiacal emission in the solar system.

TPF has two broad goals. The first is to study planetary systems, especially Earth-sized planets orbiting any of the several hundred nearest stars. The spectrometers on board TPF will analyze the infrared radiation from the planets' atmospheres and thereby determine their chemical composition. For a small number of the planets, it will search for molecular species such as ozone and methane that might indicate life. The latter observation is so difficult that TPF would require 2 weeks to observe each planet. The discovery of life on another planet is potentially one of the most important scientific advances of this century, let alone this decade, and it would have enormous philosophical implications.

TPF's second goal is to study the structure of astronomical sources at infrared wavelengths with unprecedented clarity, at an angular resolution 100 times finer than previously possible. TPF will reveal planets in the process of formation, clearing gaps in dusty protoplanetary disks; individual star formation regions in distant galaxies and in the central regions of galaxies where enormous bursts of star formation occur; and the accretion disks that feed enormous black holes in the centers of galaxies, producing quasars and related active galactic nuclei.

- **To ensure a broad science return from TPF, the committee recommends that, in planning the mission, comparable weight be given to the two broad science goals: studying planetary systems and studying the structure of astronomical sources at infrared wavelengths.**

Possible theory challenges associated with TPF are

❖ *To understand the formation and evolution of Earth-like planets and their atmospheres; and*

❖ *To understand the unique objects and processes that occur at the centers of galaxies—stellar collisions, tidal disruption of stars, supermassive black holes, accretion disks, and relativistic jets—and to understand how their interplay leads to the complex phenomena of active galactic nuclei.*

The committee notes that TPF requires the following precursor missions for technology development: ground-based OIR interferometers to demonstrate nulling and to measure the exozodiacal light in many nearby star systems; SIM to demonstrate nulling and other interferometric techniques in space; the Space Technology 3 (ST-3) mission to demonstrate formation flying; and NGST to demonstrate passive cooling, infrared detectors, large cryo-optics, and pointing, stability, and vibration control. NASA already has studies under way to determine whether the infrared emission from interplanetary dust will significantly hamper the ability of TPF to detect terrestrial planets. In addition, to ensure that TPF reaches its full scientific potential, it is important to determine prior to the start of the mission the likely probability that there will be an adequate number of Earth-sized planets for TPF to study. TPF will be a significant early step in efforts to learn more about Earth-like planets and whether they harbor life.

ULTRAVIOLET AND OPTICAL ASTRONOMY FROM SPACE

The Hubble Space Telescope has arguably had a greater impact on astronomy than any instrument since the original astronomical telescope of Galileo. Not only has it provided valuable data in virtually every area of modern astronomy, but it has also proved to be a powerful tool for inspiring popular interest in science. The committee endorses the current plans that call for HST to continue operation until the end of the decade, with reduced operating costs after completion of the final servicing mission. Other UVOIR spacecraft now in operation are the Far Ultraviolet Spectroscopy Explorer, which operates in the wavelength range from 0.09 to 0.12 μm, largely below the range of HST, and the Extreme Ultraviolet Explorer, which operates at yet shorter wavelengths. The Galaxy Evolution Explorer, which is in development, will carry out a sensitive survey of the entire sky in the UV, covering the wavelength range from 0.135 to 0.3 μm.

The committee has not recommended any new moderate or major missions for space-based UV or optical astronomy for this decade. This difficult decision was made for several reasons. First, many of the key science opportunities in UVOIR astronomy are predominantly in the infrared: Star and planet formation is best observed in that part of the spectrum because infrared radiation can penetrate the dusty medium in which the stars and planets form, and the first galaxies must be observed in the infrared because their optical and UV light is shifted into the infrared by the expansion of the universe. Second, the infrared region of the spectrum has been studied much less than the optical region, so the potential for discovery is much greater. Finally, much of the important optical astronomy can be done from the ground.

However, it is impossible to observe UV radiation from the ground, and for this decade at least, it will be impossible to carry out diffraction-limited, wide-field imaging in the optical part of the spectrum from the ground. UV observations are essential for tracing the evolution of the intergalactic gas that is too cool to emit x rays, and high-resolution UV spectroscopy is essential to study the dynamics and composition of interstellar gas. Diffraction-limited, wide-field imaging enables a search for sources too faint to be discovered from the ground. To make substantial advances on these questions beyond what can be learned from HST will require a UV-optical space telescope with a spectrometer that delivers a 100-fold increase in throughput and multiplex efficiency. To prepare the way for such a mission in the decade 2010 to 2020, the committee recommends an aggressive technology development program to develop UV detectors that are more sensitive, energy-resolving detectors such as superconducting tunnel junctions or transition edge sensors, and large, lightweight precision mirrors.

SOLAR ASTRONOMY

The internal structure and dynamics of the Sun, the resulting cyclic and random generation and dissipation of magnetic fields, and the consequent production of the solar wind are of great interest both in their own right and as nearby analogs for key astrophysical processes in more distant objects. In addition, these processes are central to understanding the effect of the Sun on Earth, and as a result they bear on the quest to determine the origin and extent of life in the universe. The study of these processes requires dedicated telescopes equipped with specialized instruments (Table 3.5).

To provide the basis for a broad advance in understanding of the magnetic and hydrodynamic processes that govern the solar surface, the committee recommends three telescopes that monitor different layers of the Sun at wavelengths that include the radio, near infrared, optical, ultraviolet, and soft x-ray. As a complement to these three telescopes, the committee notes that a small initiative to fund the expansion of the Synoptic Optical Long-term Investigations of the Sun (SOLIS) project from a one-station to a three-station network around the globe would permit nearly continuous monitoring of the solar vector magnetic field structure across the surface of the visible Sun over a long time period.

GROUND-BASED SOLAR ASTRONOMY: AST AND FASR

The Advanced Solar Telescope (AST), a 4-m-class solar telescope with adaptive optics operating from 0.3 to 35 μm, is the committee's top recommendation for solar astronomy. AST will provide a qualitative improvement over the 40-year-old Kitt Peak McMath-Pierce Telescope, which at 1.5 m is currently the largest OIR solar telescope in the world. AST will observe solar plasma processes and magnetic fields with unprecedented resolution in space and time, providing a unique opportunity to probe cosmic magnetic fields and test theories of their generation,

TABLE 3.5 Solar Telescopes

Project	Year(s)[a]	Wavelength Band[b]	Comments[c]
Ground-based Solar Telescopes			
Existing and Approved			
Dunn Solar	1969	O, IR	AO, imaging and spectroscopy
Kitt Peak McMath-Pierce	1961	O, IR	Largest currently (1.5 m)
Swedish Vacuum Tower	1985	O, IR	AO
Nobeyama Radioheliograph	1992	R	2 frequencies
New Initiatives			
AST	2009	O, IR	AO, largest (4 m)
FASR	2005	R	Multifrequency
SOLIS Expansion	2004	O	Short-term variability
Space-based Solar Telescopes			
Existing and Approved			
HESSI (U.S.)	2000 to 2003	G, X	Solar flares
SOHO (ESA, U.S.)	1995 to 2006	X, EUV, UV, O	Multipurpose
Solar-B (Japan, U.S., U.K.)	2004 to 2010	X, EUV, UV, O	Magnetic fields
STEREO	2004 to 2008	O, EUV	Two small telescopes
TRACE (U.S.)	1999 to 2004	EUV, UV	SMEX
Yohkoh (Japan, U.S., U.K.)	1991 to 2003	X	Coronal studies
New Initiative			
SDO	2006 to 2016	EUV, UV, O	Multipurpose

[a]Commencement of operations for ground-based telescopes, and scheduled years of operation for space-based telescopes.

[b]Notation as in Table 3.4, but with extreme ultraviolet (EUV) = 0.01 to 0.1 μm, x ray (X) = 0.1 to 100 keV, and gamma ray (G) = greater than 100 keV.

[c]AO, adaptive optics; SMEX, small Explorer mission.

structure, and dynamics. AST will be the first solar telescope large enough to observe structure at the fundamental length scale of 70 km, which is both the pressure scale height and the distance a photon travels before being absorbed or scattered at the solar surface. Many effects and phenomena that are observed on global scales on the Sun and other stars have their origin in physical processes that occur on this scale, and the Sun is the only star where this scale can be resolved. AST is proposed as a joint project with international partners that will be centered at the National Solar Observatory (NSO). Recent major advances in solar adaptive optics, open-air solar telescopes that provide diffraction-limited images, and large-format infrared cameras make it possible to realize AST in this decade. A theory challenge associated with AST might be

❖ *To use powerful three-dimensional magnetohydrodynamic numerical codes to model the solar activity cycle.*

The Frequency Agile Solar Radio telescope (FASR) will operate over the broad frequency range from 0.3 to 30 GHz with an angular resolution of 40 to 0.5 arcsec. It will follow the dynamic nature of solar variability at radio wavelengths and probe a range of solar surface layers from the chromosphere to the corona. Its science goals include studying transient energetic phenomena, the coronal magnetic field, and the structure of the solar atmosphere. A possible theory challenge for FASR is

❖ *To understand the dynamic transition region and corona of the Sun.*

SPACE-BASED SOLAR ASTRONOMY: SDO

The Solar Dynamics Observer (SDO) is recommended to probe more energetic processes on the Sun and to study the region below the solar surface. A successor to the extremely productive Solar and Heliospheric Observatory (SOHO), SDO is a space-based telescope in a geosynchronous orbit that will continuously monitor the Sun in the wavelength band ranging from 0.02 to 1 μm (the extreme ultraviolet to optical part of the spectrum). Many of these wavelengths are obscured by Earth's atmosphere, and therefore these space observations will provide an essential complement to the observations made by AST from the ground and to other spacecraft (see Table 3.5). SDO will help determine the origin of sunspots and solar active regions, the causes of the emergence and evolution of magnetic fields on the solar surface, the origin of coronal mass ejections and solar flares, and the connections between the interior dynamics and the activity of the solar corona. An understanding of the magnetic processes leading to solar mass ejection and flares, as well as to the generation of slow and fast solar winds, is critical to investigations of the Sun-Earth connection, including the "space weather" that has a variety of impacts on human activities. A possible theory challenge for SDO is

❖ *To model the interaction of turbulent convection and magnetic flux tubes and the interaction and reconnection of magnetic structures.*

THE HIGH-ENERGY UNIVERSE

During the period that stars are supported by the heat derived from nuclear fusion, most of the radiation they emit is in the UVOIR part of the spectrum. As in the case of the Sun, however, a small part of the radiation from these stars is in the form of high-energy photons (x rays and gamma rays) produced by magnetic phenomena at the stellar surface. When a star exhausts its nuclear fuel and it is "reborn" as a compact object—a black hole, a neutron star, or a white dwarf—much of the radiation it subsequently emits is in the form of high-energy photons. The formation of a black hole or a neutron star is generally associated with a cataclysmic event such as a supernova or possibly a gamma-ray burst. These events release an enormous amount of gravitational energy, much of which is carried away by energetic neutrinos and, to a lesser extent, by gravitational waves. All these phenomena can result in the production of energetic particles as well. The high-energy universe looks very different from the one we can see in the UVOIR part of the spectrum, and it contains clues that are vital to achieving a comprehensive understanding of the universe.

HIGH-ENERGY PHOTONS: CON-X, GLAST, VERITAS, AND EXIST

Atoms and ions emit and absorb x rays with specific wavelengths or, equivalently, specific photon energies. Gas must be hot, with a temperature exceeding hundreds of thousands of degrees, to emit x rays, but gas and dust can absorb x rays over a wide temperature range. X-ray spectroscopy therefore has the power to reveal the composition of a gas, its temperature and ionization, and (by means of the Doppler shift) its dynamics. As a result, x-ray spectroscopy can tell astronomers when and where elements were formed in the universe, how matter is distributed in clusters of galaxies and the intergalactic medium, and the dynamics of gas motions in regions around compact objects such as black holes and neutron stars. Not all x rays have energies characteristic of specific atoms or ions, but their spectrum provides clues about the conditions in the gas from which they were emitted.

Constellation-X (Con-X) is the highest-priority large project in high-energy astrophysics. Constellation-X is a system of x-ray telescopes that will go into orbit a million miles from Earth. It will measure the spectra of astronomical objects over the range from 0.25 to 40 keV with a sensitivity far greater than the existing Chandra and X-ray Multi-Mirror (XMM-Newton) missions (Figure 3.5). It achieves this dramatic improvement in sensitivity, or collecting area, by a novel design incorporating four telescopes on four separate but identical spacecraft. This design is more cost-effective than the alternative of building one much larger telescope, and also dramatically reduces the risk. Constellation-X will complement Chandra much as Keck and Gemini complement HST and as GSMT will complement NGST, by obtaining spectra of objects that the smaller telescope can just detect. It will provide an improvement in sensitivity by a factor of 20 to 300 and in spectral resolution by a factor of 3 to 10 compared with Chandra and XMM-Newton. Its timely development is now especially imperative given the tragic loss of the Japanese U.S. Astro-E mission in February 2000, which would have addressed similar goals at much lower sensitivity. Chandra, on the other hand, can make images with much higher spatial resolution and correspondingly greater positional accuracy than will Constellation-X.

Constellation-X will provide a powerful probe of the hot intergalactic medium, which may contain most of the ordinary matter in the universe. Its spectroscopic capabilities will enable it to trace the evolution of elements heavier than hydrogen and helium over cosmic time. By observing magnetized gas falling into black holes, Constellation-X will probe the properties of spacetime near a spinning black hole, thereby testing strong-field general relativity. Neutron stars, by contrast, have a solid surface and emit x rays in a rich variety of spectral lines. By analyzing the wavelengths, widths, and relative strengths of these lines, it is possible to determine the mass and radius of the star, just as one can with a star like the Sun. This information will ultimately provide a constraint on the behavior of quarks and gluons, the fundamental building blocks of matter, under conditions that cannot be reproduced by particle accelerators.

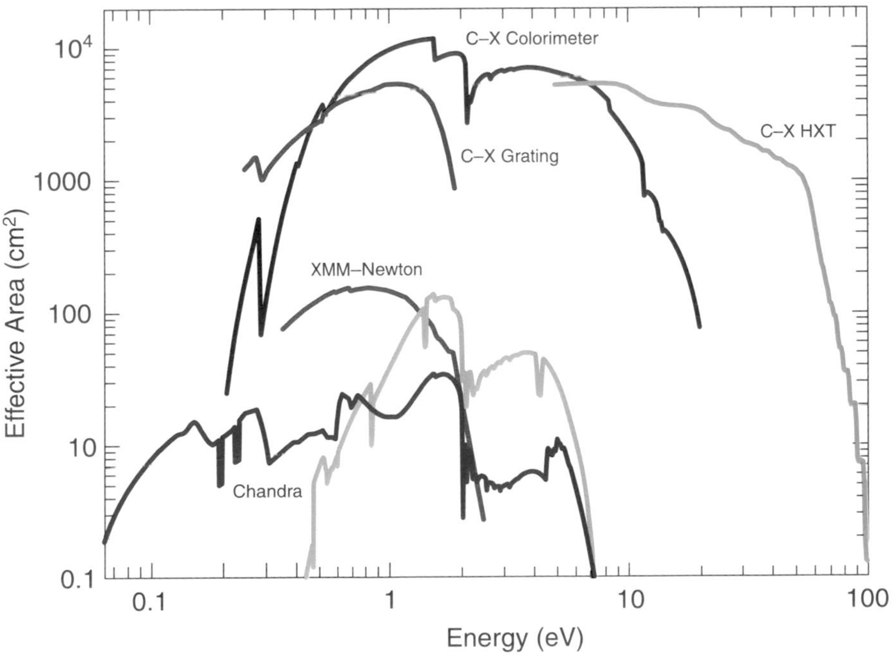

FIGURE 3.5 Comparison of the effective collecting area (equivalent to sensitivity) for x-ray spectroscopy as a function of energy between Constellation-X, Chandra, and XMM-Newton. Note that Constellation-X uses three separate devices to achieve a uniformly large collecting area from roughly 0.25 to 40 keV. The effective areas of the two transmission grating spectrometers on board Chandra are shown in dark blue and light blue. Courtesy of NASA and the Smithsonian Astrophysical Observatory.

Possible theory challenges for Constellation-X are

❖ *To develop accurate, documented general-purpose computer codes to model general-relativistic magnetohydrodynamics; and*

❖ *To understand how the distribution of the elements has evolved over cosmic time.*

GLAST and VERITAS. Gamma rays are photons even more energetic than x rays, with energies above a few hundred keV. Some gamma rays have energies characteristic of the nucleus of the atom from which they were emitted; others are produced by electrons that are more energetic than those that produce x rays. Relatively low energy gamma rays must be observed from space, but gamma rays with photon energies above about 50 GeV can be detected by ground-based telescopes that observe the light generated when the gamma rays strike molecules in the atmosphere. Currently the Energetic Gamma Ray Experiment (EGRET) on the spaceborne Compton Gamma Ray Observatory (CGRO) scans the universe for 0.1- to 10-GeV gamma rays, and the ground-based Whipple Observatory looks for 200- to 3,000-GeV gamma rays. Two specialized Explorer missions, the High Energy Transient Explorer (HETE-2) and Swift, are poised to locate gamma-ray bursts precisely and enable rapid follow-up observations at other wavelengths. Integral, a European mission to be launched in 2001, will measure spectra at photon energies of 20 keV to 10 MeV. None of the existing or planned observatories have the sensitivity to detect very many sources—they are what astronomers call "starved for photons." The committee recommends two missions to cover the broad band of gamma-ray energies with high sensitivity: the Gamma-ray Large Area Space Telescope (GLAST), for photon energies from 10 MeV to 300 GeV, and the ground-based VERITAS project to cover 50 GeV to 10,000 GeV. For the first time, there will be an overlap in the photon energies detectable from ground and space. GLAST will view a much wider number and variety of sources than VERITAS and is the highest-priority moderate space mission. Figure 3.6 compares the relative sensitivities of GLAST and VERITAS with those of their predecessors, EGRET and Whipple. Improvement by a factor of 10 to 30 in sensitivity means that the number of sources that can be studied improves by a factor of 30 to 150. This increase in sensitivity opens up significant opportunities for new discoveries, particularly when combined with the improved angular resolution these projects provide. GLAST and VERITAS will both address important issues concerning jets from active galactic nuclei, the acceleration of cosmic rays, and the nature of gamma-ray bursts.

A possible theory challenge for both GLAST and VERITAS is

❖ *To model the dynamics of and the emission from the relativistic jets that emanate from the vicinity of central black holes in active galaxies.*

The Energetic X-ray Imaging Survey Telescope (EXIST) is a specialized survey telescope recommended to study the sky at lower energies (5 to 600 keV). It will perform a survey 1,000 times more sensitive than the previous survey in this energy range, which was done by the High Energy Astronomical Observatory (HEAO-1). Figure 3.6 shows how the EXIST survey will complement those by the Roentgen Satellite (ROSAT),

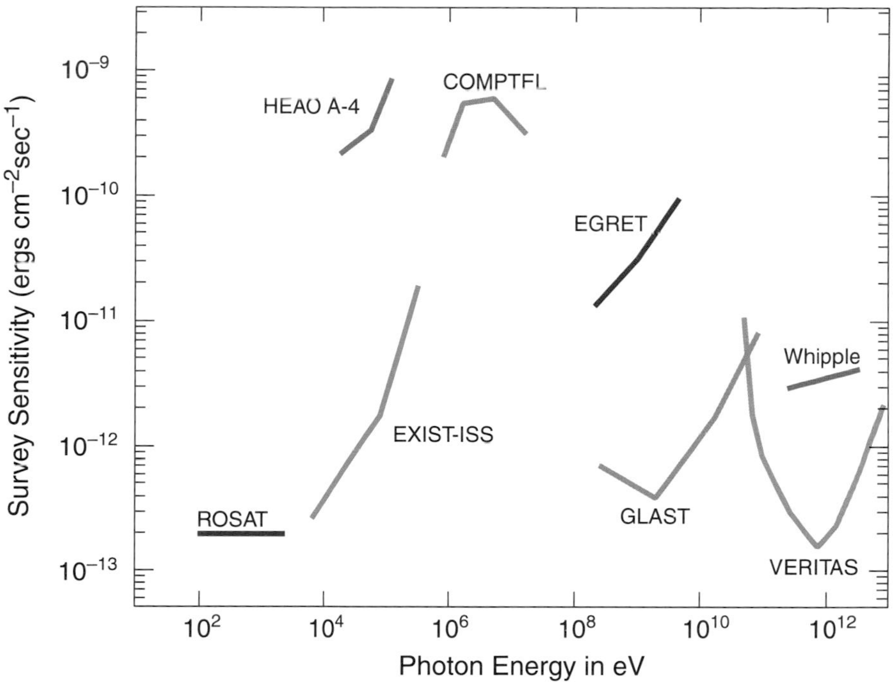

FIGURE 3.6 Limiting fluxes for existing and projected x-ray and gamma-ray surveys as a function of energy. The fluxes assume an energy band equal to the energy, i.e., a broadband measurement. As in Figure 3.2, the better-performing instruments appear lower in the figure. Note the large gains in sensitivity for GLAST, EXIST, and VERITAS. COMPTEL and EGRET are instruments on the Compton Gamma Ray Observatory. Courtesy of L. Bildsten (University of California, Santa Barbara), and NASA.

CGRO, GLAST, and VERITAS. EXIST is a space station-attached telescope with a spectral resolving power of 100 and the ability to locate bright sources to about 30 arcsec. EXIST will survey the entire sky in every 90-minute orbit, which allows the study of the highly time-variable sources that characterize the x-ray sky. It will carry the study of gamma-ray bursts to lower energies and will be able to study the low-power gamma-ray bursts that appear to be associated with supernovae. Because energetic x rays are so penetrating, EXIST can discover supernovae embedded in molecular clouds and the luminous matter accreting onto supermassive black holes in the centers of galaxies that are obscured at lower photon energies by surrounding gas and dust. Finally, with its ability to perform high-energy-resolution, hard-x-ray spectroscopy of neutron stars, EXIST will enable astrophysicists to study how radiation interacts with magnetic fields that can be a million to a billion times stronger than can be sustained in the laboratory.

A possible theory challenge for EXIST is

❖ *To solve the mystery of gamma-ray bursts.*

GRAVITATIONAL RADIATION: LISA

The direct measurement of gravitational waves from astrophysical sources will open new investigations in both astrophysics and the physics of strong gravitational fields. Gravitational waves can probe the dense inner regions of astrophysical systems that are opaque to photons. The committee recommends the Laser Interferometer Space Antenna (LISA), a joint mission between the United States and the European Space Agency, to pioneer the study of low-frequency (periods of 10 to 10,000 seconds) gravitational waves from binary star systems in our galaxy and the coalescence of supermassive black holes. A mission of this type was recommended in the physics survey report *Gravitational Physics: Exploring the Structure of Space and Time* (NRC, 1999). LISA will complement the ground-based Laser Interferometer Gravitational-wave Observatory (LIGO), which is designed to study the much higher frequency gravitational waves from the coalescence of neutron stars and stellar mass black holes, as well as the core collapse of supernovae. The detection of low-frequency gravitational waves requires a space system with detectors several million miles apart whose separation is monitored with exquisite accuracy, to a precision a thousand times smaller than the size of an atom. Although much progress has been made in the technology for such a mission, including the ground-based laser interferometry for LIGO and the shielding of the reference mass detectors by the space-based Triad and Gravity Probe B programs, a dedicated technology mission in space is envisioned as a precursor to LISA. Three technical areas could benefit from an integrated test on such a precursor: the inertial reference mass, precision thrusters, and high-precision interferometry.

A theory challenge appropriate for LISA would be

❖ *To compute the expected gravitational waveforms from black hole mergers.*

PARTICLE ASTROPHYSICS

Although most of what we know about the high-energy universe comes from photons, crucial information is also carried by energetic particles—cosmic rays, neutrinos, and, possibly, exotic particles that could constitute much of the mass of the universe, the so-called dark matter. Cosmic rays are relativistic particles accelerated by supernova shock waves and other energetic phenomena. They play an important role in the ionization, heating, and pressurizing of the interstellar medium and in the production of high-energy photons. The typical cosmic ray has an energy of motion comparable to the energy associated with its rest mass, but the most energetic cosmic rays have energies nearly a trillion times greater. The nature and origin of these ultrahigh-energy cosmic rays are not understood. The Southern Hemisphere Pierre Auger Observatory and the high-resolution Fly's Eye are two ground-based projects that will soon be under way to detect and characterize these ultrahigh-energy cosmic rays. The composition of lower-energy cosmic rays is being studied by the Advanced Composition Explorer, whereas the Antimatter-Matter Spectrometer will search for the presence of antimatter in the cosmic rays.

The committee notes that a proposed small space mission, the Advanced Cosmic-ray Composition Experiment for the Space Station (ACCESS), shows great promise in being able to characterize the mechanism of cosmic-ray acceleration with far greater precision than heretofore possible. A possible theory challenge for ACCESS is

❖ *To study the acceleration and propagation of relativistic particles in astrophysics in order to enable accurate comparison between theory and ACCESS observations.*

Neutrinos are produced by nuclear reactions in the interior of stars like the Sun, in supernova explosions, and possibly in gamma-ray bursts and in the regions around supermassive black holes. Most existing neutrino detectors are designed to study the relatively low energy neutrinos from the Sun and are focused primarily on studying the physics of neutrinos; as such, they lie outside the purview of this report. Several projects are under way in Europe and in the United States to search for much-higher-energy neutrinos. The U.S. project, AMANDA, uses a huge volume of subsurface ice at the South Pole to detect neutrinos; a much larger follow-on experiment, Ice Cube, has been proposed.

Searches for exotic dark matter particles are also under way or about to begin. In the United States, the Axion experiment is just beginning its search for the "axion" particle, whereas the Cryogenic Dark Matter Search II project promises to be the most sensitive experiment yet to search for the "WIMP" or neutralino particle. These experiments are crucial for astrophysics as well as physics, since our understanding of the universe will be woefully incomplete if we do not comprehend the nature of the dominant form of matter in the universe.

A characteristic of all the projects in particle astrophysics described here, with the exception of ACCESS, is that they are primarily experiments rather than observatories. The nature of experiments means that it is difficult, and likely wasteful, to plan a follow-on experiment without knowing the outcome of the original one. The committee therefore concludes that while particle astrophysics is an exciting and potentially revolutionary field, the decision on whether to proceed with initiatives such as the Northern Hemisphere Auger project/Telescope Array complex and Ice Cube should await the initial outcome of their precursors.

THE RADIO UNIVERSE

Radio waves provide a window onto the origins of the universe, galaxies, stars, and planets that is both unique and complementary to that at other wavelengths. Relic radiation from the Big Bang has been shifted in wavelength by the expansion of the universe to the radio regime, and detected as the cosmic microwave background. Perturbations in this background caused by intervening hot gas in galactic clusters permit radio astronomers to locate these clusters and to view the large-scale structure of the universe. Relativistic particles, spiraling in magnetic fields, emit radio photons and have provided astronomers with the first views of the enormous jets emanating from the vicinities of black holes at the centers of galaxies as well as the high-energy particles accelerated by supernovae. Radio waves offer a clear view of the earliest stages of star and planet formation, which is obscured at many other wavelengths by the surrounding clouds of gas and dust. Radio waves also penetrate Earth's atmosphere, so that nearly all radio telescopes are ground-based.

CENTIMETER-WAVELENGTH ASTRONOMY: EVLA, SKA, AND ARISE

Table 3.6 lists the major centimeter-wavelength observatories accessible to U.S. astronomers. For studying complex or weak, extended sources or problems that require frequency agility or large collecting area, the Green Bank Telescope (GBT) and the newly upgraded Arecibo telescope are unparalleled. The world's largest steerable antenna, the GBT uses a unique design and active surface control. In addition to being the world's largest filled aperture, the Arecibo dish can undertake

TABLE 3.6 Large Centimeter-Wave Radio Telescopes with Open Access

Project	Nations Involved	Aperture[a]	Wavelength Range (cm)	Angular Resolution[b] (arcsec)
Existing and Approved				
ATCA (ATNF)	Australia	6×22 m	0.3 to 20	0.1
Parkes (ATNF)	Australia	1×64 m	1.3 to 90	50
Arecibo	United States	1×300 m	6 to 90	60
Effelsberg	Germany	1×100 m	0.4 to 30	10
GBT	United States	1×100 m	0.3 to 150	10
GMRT	India	30×45 m	21 to 300	2
HALCA	Japan, United States	1×8 m	6 to 20	10^{-3}
1HT	United States	500×5 m	3 to 30	3
MERLIN	United Kingdom	$6 \times (25\text{-}76)$ m	1.3 to 200	0.01
Nançay	France	$1 \times (35$ m $\times 300$ m)	9 to 21	100
VLBA	United States	10×25 m	0.4 to 90	10^{-4}
VLA	United States	27×25 m	0.7 to 400	0.04
Westerbork	Netherlands	14×25 m	6 to 150	4
New Initiatives				
EVLA	United States	37×25 m	0.7 to 400	0.007
ARISE	United States	1×25 m	0.3 to 3	10^{-5}
LOFAR	Netherlands, United States	10^6 m^2	200 to 1,000	1
Technology Development				
SKA	International	10^6 m^2		

[a]Notation: 6×22 m denotes six dishes, each with a diameter of 22 m. Nançay is a single dish but has a noncircular shape.

[b]The angular resolution shown for HALCA and for ARISE reflects their use in combination with ground-based dishes in an interferometric array.

radar remote sensing of solar system objects as well as passive radio astronomy. For compact sources, however, it is necessary to use interferometry, a technique pioneered by radio astronomers in which radio signals detected at antennas separated by distances of up to thousands of miles are combined to form an image of the source. The highest resolutions are achievable with the Very Long Baseline Array (VLBA), used to image radio sources associated with quasars, active galactic nuclei (AGNs), molecular masers, and other compact sources. For larger radio sources, however, the Very Large Array (VLA) is the instrument of choice. It provides spatial resolutions of 0.1 to 1 arcsec, comparable to those of the large ground-based optical telescopes. The VLA is the most powerful and productive centimeter-wave telescope in the world, despite the fact that its instrumentation is 25 to 30 years old.

The Expanded VLA (EVLA) is the second priority among major, ground-based projects. The EVLA will have 10 times the sensitivity and angular resolution and 1,000 times the spectroscopic capability of the VLA. The first stage of the expansion will replace instruments, computers, and software and install wideband fiber-optics data links. In the second stage, up to eight new antennas will be sited within 250 km of the VLA and connected via fiber-optics links. The resulting angular resolution of 0.01 to 0.1 arcsec will be comparable to that of ALMA and of NGST, facilitating multi-wavelength studies. These new antennas will also enhance the field of view and sensitivity of the VLBA when the two systems are used together. The committee also notes that a complementary small project, the Low Frequency Array (LOFAR), would extend wavelength coverage to 20 m and provide improvement by a factor of 100 to 1,000 in sensitivity and resolution over existing instruments at these wavelengths. The overall dimension of LOFAR would be several hundred kilometers, possibly using the VLA site as the primary location.

The high angular resolution and sensitivity of the EVLA, combined with the penetrating power of centimeter-wave radio waves, will enable the detailed study of nearby protostars and protoplanetary disks as well as the active nuclei of distant galaxies. The EVLA will produce images of protogalaxies with sufficient detail to determine whether AGN activity associated with a supermassive black hole precedes, is contemporaneous with, or follows bursts of star formation in galactic nuclei.

A possible theory challenge for the EVLA is

❖ *To understand the roles of star formation and supermassive black holes in powering luminous active galactic nuclei.*

The Square Kilometer Array Technology Development. The EVLA, VLBA, LOFAR, and One Hectare Telescope (1HT), a privately funded array to be used in part for the search for extraterrestrial intelligence (SETI) project, form the foundation of ground-based interferometric centimeter-wave astronomy for this decade. To study how the first galaxies condensed out of vast clouds of atomic hydrogen, a substantially larger radio telescope is needed: the Square Kilometer Array (SKA), with 1 million square meters of collecting area. (By comparison, the EVLA has some 13,000 square meters of collecting area.) The extraordinary sensitivity of SKA will also revolutionize fields of study newly accessible to centimeter-wave astronomy, including the study of jets and the disks of protostars, the measurement of magnetic fields in collapsing clouds, and the study of the distribution of dark matter on the largest scales by means of weak gravitational lensing. The SKA is a major international project that may start in the decade 2010 to 2020. The committee recommends a coherent development program over the current decade to develop the technology that will enable the science objectives to be met at a reasonable cost.

A possible theory challenge for the SKA development is

❖ *To understand the formation of the first generation of stars and their effect on reionizing the universe.*

The Advanced Radio Interferometry between Space and Earth (ARISE) mission is a 25-m-class space antenna to be linked with the ground-based VLBA. It is recommended as a means of achieving the highest spatial resolution for bright sources such as jets emanating from near supermassive black holes in galactic nuclei. ARISE will operate at wavelengths as short as about 3 mm. Its elliptical orbit will reach up to 50,000 km from Earth, giving an angular resolution six times better than that obtained with the VLBA. It will be an order of magnitude more sensitive than the Japanese HALCA space interferometric system that is currently in operation.

A theory challenge for ARISE might be

❖ *To understand maser emission and other physical processes in nonrelativistic accretion disks in our own and other galaxies.*

MILLIMETER- AND SUBMILLIMETER-WAVE ASTRONOMY: CARMA AND SPST

Table 3.7 lists the major ground-based millimeter- and submillimeter-wave telescopes in the world. The Atacama Large Millimeter Array (ALMA), an international collaboration among the United States, Canada, Europe, and possibly Japan, will be by far the most powerful of these telescopes when it becomes operational. Situated at a high, dry site in Chile, ALMA will dominate millimeter-wave observations of the southern sky. Under another name (the Millimeter Array), ALMA was the first-ranked radio project a decade ago (NRC, 1991); the committee reaffirms support for this project. The high spatial resolution (as fine as 0.01 arcsec) and sensitivity of ALMA will allow unprecedented views of diverse astronomical phenomena ranging from comets and Kuiper Belt objects in the solar system, to planet-forming disks in nearby regions of star formation, to the structure of the interstellar medium in distant galaxies.

The Combined Array for Research in Millimeter-wave Astronomy (CARMA). The committee recommends support for the construction of a Northern Hemisphere array, somewhat different in design, that will complement ALMA. CARMA would combine nine of the current 6-m Berkeley-Illinois-Maryland Association (BIMA) antennas, the six 10.4-m Owens Valley Radio Observatory (OVRO) dishes, and ten new 2.5-m antennas at a higher and better site in California. The resultant hybrid array would offer unique imaging capabilities to study structure on all scales with particular sensitivity to low-surface-brightness, extended emission. CARMA will be a powerful tool for studying the chemistry, dynamics, and structure in star-forming regions as well as for mapping the deviations in the cosmic microwave background caused by the hot gas in clusters of galaxies. As a project, CARMA is run by a university consortium and is largely funded by nonfederal sources but will provide significant access to the entire astronomical community. It could be undertaken immediately, fostering the training of students and the U.S. capability in millimeter-wave interferometry at the start of the ALMA era and beyond.

The South Pole Submillimeter-wave Telescope (SPST). For its combination of low opacity and stable seeing, the South Pole is the best site in the world for ground-based observations at submillimeter wavelengths. To take advantage of the opportunities offered by this site, the committee recommends the construction there of a 7- to 10-m-class filled-aperture submillimeter-wave telescope. Such a telescope should be equipped to survey the sky so that it can identify sources such as primordial galaxies, study the distortion of the cosmic microwave background caused by clusters of galaxies, and survey the dusty universe. Its survey capability will make the SPST an important complement to ALMA.

A possible theory challenge for CARMA and SPST is

❖ *To understand the dynamical and chemical evolution of molecular clouds in galaxies.*

THE COSMIC MICROWAVE BACKGROUND RADIATION

The cosmic microwave background radiation was emitted early in the history of the universe, before stars and galaxies formed. The radiation was emitted with a spectrum dominated by optical and near-infrared radiation, but the expansion of the universe has increased its wavelengths by a factor of about 1,000, so that it is now concentrated in the millimeter and submillimeter parts of the spectrum. When the radiation was emitted, the universe was almost, but not quite, perfectly homogeneous. The small inhomogeneities present at that time were the seeds of the formation of galaxies, clusters of galaxies, and larger structures in the universe. The Cosmic Background Explorer (COBE) was flown in the past decade and made the most precise measurements at that time of the background radiation and the tiny variations in its intensity over the sky (about 1 part in 100,000). Since then, ground- and balloon-based experiments have probed the microwave background on smaller angular scales than did COBE. Recent balloon observations have shown that the total density of matter and energy is just what is needed to make the geometry of the universe flat (see Chapter 2).

Measurements of the microwave background are the primary means available for probing the large-scale structure of the early universe, and as such they are essential for addressing one of the primary science goals for this decade. Scientists' knowledge of the microwave background will be transformed by ongoing ground-based experiments and by the upcoming Microwave Anisotropy Probe (MAP) MIDEX (mid-size Explorer) mission. The Planck Surveyor, a European-led mission with significant U.S. involvement that is planned for launch in 2007, will study smaller-scale fluctuations in the background. Future microwave background experiments, such as measuring the polarization, are of great importance, but the committee recommends that prioritization of such

TABLE 3.7 Ground-Based Millimeter- and Submillimeter-wave Telescopes

Project	Nations Involved	Wavelength Band	Aperture[a]	Angular Resolution[b] (arcsec)
Existing and Approved Interferometers				
ALMA	United States, Europe, Japan	Millimeter	64 × 12 m	$0.003\left(\frac{\lambda}{0.3\text{ mm}}\right)$
BIMA	United States[c]	Millimeter	10 × 6 m	$0.2\left(\frac{\lambda}{1\text{ mm}}\right)$
IRAM	Europe	Millimeter	5 × 15 m	$0.6\left(\frac{\lambda}{1.5\text{ mm}}\right)$
Nobeyama	Japan	Millimeter	6 × 10 m	$1.5\left(\frac{\lambda}{3\text{ mm}}\right)$
OVRO	United States[c]	Millimeter	6 × 10.4 m	$0.5\left(\frac{\lambda}{1.5\text{ mm}}\right)$
SMA	United States, Taiwan	Submillimeter	8 × 6 m	$0.1\left(\frac{\lambda}{0.3\text{ mm}}\right)$
New Initiative				
CARMA	United States[c]	Millimeter	25 × (2.5 to 10 m)	$0.1\left(\frac{\lambda}{1\text{ mm}}\right)$
Existing and Approved Single Dish				
CSO	United States[c]	Submillimeter	10.4 m	$7\left(\frac{\lambda}{0.3\text{ mm}}\right)$
FCRAO	United States[c]	Millimeter	14 m	$50\left(\frac{\lambda}{3\text{ mm}}\right)$
Hertz (SMT)	Germany, United States[c]	Submillimeter	10 m	$7\left(\frac{\lambda}{0.3\text{ mm}}\right)$
IRAM	Europe	Millimeter	30 m	$25\left(\frac{\lambda}{3\text{ mm}}\right)$

TABLE 3.7 Continued

Project	Nations Involved	Wavelength Band	Aperture[a]	Angular Resolution[b] (arcsec)
Existing and Approved Single Dish (continued)				
JCMT	United Kingdom, Netherlands, Canada	Submillimeter	15 m	$5\left(\frac{\lambda}{0.3\text{ mm}}\right)$
LMT	United States,[c] Mexico	Millimeter	50 m	$7\left(\frac{\lambda}{1.5\text{ mm}}\right)$
Nobeyama	Japan	Millimeter	45 m	$16\left(\frac{\lambda}{3\text{ mm}}\right)$
New Initiative				
SPST	United States	Submillimeter	7 to 10 m	$7\left(\frac{\lambda}{0.3\text{ mm}}\right)$

[a]Notation: 64×12 m denotes 64 antennas with 12-m diameters.
[b]The wavelength is scaled to the shortest operating wavelength; the number in front of the parenthesis is the angular resolution at that wavelength in arcseconds.
[c]University or private facility in the United States.

experiments await the results from MAP (assuming it is successful) and the ongoing suite of ground-based and balloon projects.

THE SEARCH FOR EXTRATERRESTRIAL INTELLIGENCE

Are we alone in the universe? Finding evidence for intelligence elsewhere would have a profound effect on humanity. Searching for evidence for extraterrestrial life of any form is technically very demanding, but, as indicated in the discussion of TPF above, there is a clear approach for doing so. The search for extraterrestrial intelligence is far more speculative because researchers do not know what to search for. Radio astronomers have taken the lead in addressing this challenging problem, and SETI programs are under way at many radio telescopes around the world. This committee, like previous survey committees, believes that the speculative nature of SETI research demands continued development of innovative technology and approaches, which need not be restricted to radio wavelengths. The privately funded 1HT, which will be the first radio telescope built specifically for SETI research, is a good example of such an innovative approach, and it will pioneer new radio techniques that could be used in the SKA.

THE NATIONAL VIRTUAL OBSERVATORY AND OTHER HIGH-LEVERAGE, SMALL INITIATIVES

The National Virtual Observatory (NVO). As the new millennium begins, astronomy faces a revolution in data collection, storage, analysis, and interpretation of large data sets. Data are already streaming in from surveys such as the Two Micron All Sky Survey and the Sloan Digital Sky Survey, which are providing maps of the sky at infrared and optical wavelengths, respectively. The LSST, which will survey the sky every 3 days, will add a third dimension, time, to the data. What is needed is to make the data from all these surveys available to astronomers, educators, and the public so that they can view images of the evolving sky at any wavelength (or color) surveyed by astronomical telescopes. Each day, trillions of bits of information from telescopes will have to be rapidly archived and made available for viewing and analysis. The NVO is designed to enable this, and it is the committee's highest priority for small projects.

The NVO will link the major astronomical data assets into an integrated, but virtual, system to allow automated multiwavelength search and discovery among all cataloged astronomical objects. The computers used in the NVO would be distributed across the country, but high-speed networks would link them into a unified system. The NVO not only would archive the data but also would provide advanced analysis services for the astronomical community, create data standards and tools for mining data, and provide a link between the exciting astronomical data and the educational system in the United States. The opportunities for communicating the most recent discoveries in the dynamic sky into classrooms and homes worldwide give the NVO the potential of becoming a powerful tool for increasing the general public's science literacy.

Theoretical Astrophysics. All the major and moderate programs recommended in this report are aimed at making substantial advances in astronomers' ability to observe the universe. However, these data would have little significance in the absence of a theory to interpret them. Theorists play a major role in defining the intellectual frontier of astronomy and astrophysics, in developing the models that quantitatively relate observational data to the underlying physics and chemistry, and in synthesizing a world view that is accessible to the general public. In view of the explosion in the rate of astronomical discoveries, the committee believes that the resources recently devoted to theory are not adequate for an optimized program, and it therefore recommends three small initiatives to redress this imbalance: (1) Theory challenges tied to major and moderate projects. These challenges should be administered as a competitive grants program that is budgeted and programmed as an integral part of its associated project. Examples of possible theory challenges are given above. Each challenge should identify a theoretical problem that is ripe for progress, relevant to the planning and design of the mission, and essential to the interpretation and understanding of its results in the broadest context. Funding for this program might typically amount to 2 to 3 percent of the cost of the project, although this amount could vary substantially depending on the nature of the project. (2) The National Astrophysical Theory Postdoctoral Program modeled on the highly successful Hubble postdoctoral program. Currently, grants to individual theorists rarely have the funding or the longevity to support postdoctoral fellows. A national postdoctoral program in theoretical astrophysics will support innovative research by the new generation, foster their intellectual development, and encourage ethnic and gender diversity. This program should be supported jointly by NASA and NSF to provide 10 new 3-year postdoctoral positions each year. (3) Augmentation of the Astrophysics Theory Program at NASA. This program has been highly successful, but the report *Federal Funding of Astronomical Research* (NRC, 2000) documents the difficulties facing it—the 3-fold oversubscription in 1987 increased to oversubscription by a factor of 4.8 in 1997, and the overall funding for theory declined by about 20 percent from 1990 to 1999. The committee recommends that the Astrophysics Theory Program be augmented by $3 million per year to bring it into better balance with the vigorous experimental and observational programs at NASA. More details on these recommendations can be found in Chapter 1 of the *Panel Reports* (NRC, 2001). *These three programs should not divert funds from existing grants programs for broadly based theory.*

Laboratory Astrophysics. Existing missions and facilities are returning spectroscopic data of unprecedented breadth and detail, yet in many cases these data cannot be interpreted because the underlying atomic and molecular properties are unknown. Even in the Sun, almost 40 percent of the coronal lines observed by SOHO in the range from 50 to 160 nm are unidentified; for those lines that are identified, the wavelengths and excitation cross sections may be so poorly known that quantitative interpretation of the data is impossible. Existing missions such as Chandra and XMM-Newton and proposed missions such as NGST, Constellation-X, and SAFIR will obtain spectra of much more exotic objects where the uncertainties are much greater. In addition to atomic and molecular physics, the properties of irradiated ices, refractory grains, and fluids at high energy densities also require study. At present, support from both NASA and NSF for these types of investigations is extremely limited. The committee recommends a significant increase in support for these areas, primarily from NASA, since the largest residual uncertainties in the atomic and molecular databases pertain to transitions that lie primarily in wavelength bands accessible only from space. Specifically, the committee recommends a dedicated NASA grants program for laboratory astrophysics funded at the level of $4 million per year ($40 million for the decade), which should enable roughly 8 to 10 significant experimental programs that collectively cover most of the spectrum. The committee also recommends a smaller level of support from NSF ($500,000 per year or $5 million for the decade) that would be directed at computational atomic and molecular physics and database development.

In addition to the above recommended initiative for laboratory astrophysics, the committee notes several proposed programs that will provide significant benefit to astrophysics. Direct laboratory simulation of magnetic reconnection in plasmas and of radiative-hydrodynamical instabilities in supernovae and supernova remnants have the potential to be of great benefit in interpreting astronomical observations, particularly when coupled with computational modeling of the same phenomena. Better knowledge of key nuclear reaction rates is essential for advancing scientists' understanding of the late stages of stellar evolution, of supernovae, and of Big Bang nucleosynthesis. Finally, experiments at the Relativistic Heavy Ion Collider and the Large Hadron Collider will determine the properties of the quark-gluon plasmas that existed a few microseconds after the Big Bang.

Ultralong-Duration Balloons (ULDBs). NASA's Explorer program has been extremely successful because it provides frequent access to space for innovative projects across the entire field of astrophysics. In many cases, most of the advantage gained by going into a low Earth orbit could be achieved at far less cost with balloons capable of remaining aloft for periods of 100 days or more. Examples of the kind of science that can be carried out with balloons include searches for planets with a coronagraph on a diffraction-limited telescope a few meters in diameter, imaging convective flows and magnetic fields in the Sun's photosphere with a large solar telescope, extragalactic observations with a moderate-sized far-infrared telescope, or all-sky surveys at hard x-ray wavelengths. The top of the stratosphere is far superior to terrestrial sites and enables a wide range of small-mission science at wavelengths not transmitted to Earth.

The committee recommends that NASA invest the necessary resources (estimated to be about $35 million) to develop steerable ULDBs, and that use of ULDBs be allowed as an alternative to spacecraft (where warranted) in all Explorer programs.

4

Benefits to the Nation from Astronomy

INTRODUCTION

Astronomical discoveries of the past decade—images of the hot universe at an epoch before the first galaxies and stars emerged, of other solar systems beginning to take form, of planetary systems beyond our own—have captured the imagination of scientists and citizens alike. These startling advances are the result not only of the collective creative efforts of scientists and engineers throughout the United States and around the world, but also of the generous investments in astronomy over much of the past 50 years by federal and state governments, foundations, and individuals.

In the decades ahead, the pace of discovery—remarkable as it has been over the past—will accelerate. Astronomers stand poised to examine the epoch when galaxies similar to our Milky Way first took form, to image Earth-like planets beyond our solar system, and to learn whether some show evidence of life. To take these next steps will require significant investments of both imagination and public resources.

Because the magnitude of these investments will be large, it is fair to ask why astronomical research should merit such support. Perhaps the most persuasive, but least quantifiable, justifications lie in the importance American society has always attached to exploring new frontiers, and in the deep human desire to understand how we came to be, the kind of universe we live in, whether we are alone, and what our ultimate fate will be. Exploring frontiers of unimaginable mystery and beauty, astronomy speaks compellingly to these fundamental questions.

As researchers, astronomers experience the excitement of discovery most vividly and are the first to glimpse new answers to ancient questions. As a community of citizens fortunate to live in a society that supports them generously, astronomers believe strongly that "from those to whom much is given, much is asked." It is in that spirit that the committee offers below an accounting of astronomy's more tangible contributions to broader societal goals.

THE ROLE OF ASTRONOMY IN PUBLIC SCIENCE EDUCATION

Astronomers' most significant contribution to society lies in the area of science education, broadly conceived to include (1) raising public awareness of science, (2) conveying scientific concepts to students at all levels and to their teachers, and (3) contributing to educating a technically capable and aware citizenry. Astronomy is relevant to each of these goals, and it can act as a pathfinder in stimulating people's interest in all of science.

THE RELEVANCE OF ASTRONOMY

Astronomy excites the imagination. The beauty of the night sky and its rhythms are at once stunning and compelling. The boldness of our collective efforts to comprehend the universe inspires us, while the dimensions of space and time humble us. Astronomy encompasses the full range of natural phenomena—from the physics of invisible elementary particles, to the nature of space and time, to biology—thus providing a powerful framework for illustrating the unity of natural phenomena and the evolution of scientific paradigms to explain them. In combination, these qualities make astronomy a valuable tool for raising pubic awareness of science, and for introducing scientific concepts and the process of scientific thinking to students at all levels. A few reminders serve to illustrate the potential of astronomy to advance public science education goals.

Astronomy is all around us. Just look up! Who has not looked at the night sky and wondered at the panoply of stars there? We are all aware of the motion of the Sun through the sky during the day and the changing phases of the Moon at night. The motions of astronomical objects determine the day-night cycle, the seasons of the year, the tides, the timing of eclipses, and the visibility of comets and meteor showers. Easily observed astronomical events have formed the basis for time keeping, navigation, and myths or sagas in cultures around the world.

Much of astronomy is visual and can be appreciated for its aesthetic appeal as well as its illustrative power. Images of deep-sky objects convey the beauty of the universe, even to those who are too young to understand their context or implications.

Astronomy is a participatory science. Many nonscientists have astronomy as a lifelong avocation. Astronomy is one of the few sciences in which amateurs by the tens of thousands have formed active organizations (e.g., the Planetary Society, with membership exceeding 130,000), and many amateurs make significant scientific contributions to such fields as the monitoring of variable stars and measuring positions of moving objects. Telescope and magazine sales suggest that nearly 300,000 citizens take some active interest in amateur astronomy. The American

Astronomical Society has formed a working group to foster partnering between professional and amateur astronomers. Many amateurs freely share their excitement about science with local teachers and students through such programs as Project ASTRO, which links astronomers with 4th through 9th grade teachers and classes in 10 sites around the country.

Astronomy offers the possibility of discovery. The chance to find a never-before-seen supernova, nova, comet, or asteroid is very exciting, especially to nonprofessionals. Both the distribution of astronomical data and software via the Internet and the ready availability of sophisticated imaging devices on moderate-cost small telescopes enable amateur astronomers to play an active and growing role in discovering new objects, searching for transient and variable objects, and monitoring them.

Astronomy inspires work in the arts. From poetry and music to science fiction books and films, the ideas and discoveries of modern astronomy serve as inspiration for artists, for youngsters, and for the public at large. In the process, the works inspired by astronomy can serve as goodwill ambassadors for the value and excitement of physical science to many in society who do not otherwise come into contact with the sciences.

CONVEYING ASTRONOMY TO THE PUBLIC

Statistics confirm the widespread interest in astronomy.

Planetariums and observatories are popular visitor destinations. There are approximately 1,100 planetariums in North America. About 30 percent of these serve school groups only, while about 70 percent do both school and public shows. Approximately 28 million visits are made to the planetariums in the United States each year. For many school children from urban areas, such a visit may be their only introduction to a dark night sky and to the wonders of the universe.

Observatory visitor centers are similarly popular. They provide a place where families learn about science together. For example, the seven observatories that belong to the Southwestern Consortium of Observatories for Public Education (McDonald, the National Solar Observatory at Sacramento Peak, Kitt Peak National Observatory, the Very Large Array (VLA), Lowell Observatory, Whipple Observatory, and Apache Point), collectively host more than 500,000 visitors annually and reach more than 4,000 teachers through workshops. The new Visitor Center at Arecibo in Puerto Rico hosts an average of 120,000 visitors each year. Most science museums have sections on astronomy and hold weekend, evening, and summer programs on astronomical sciences.

Astronomy serves as an introduction to science for nearly 10 percent of all college students—more than 200,000 each year, nationwide. For many, astronomy will be the only science course they will ever take. To examine and improve the effectiveness of teaching science via introductory astronomy courses—many of which are offered at community colleges and small colleges without extensive research programs—the Astronomical Society of the Pacific and the American Astronomical Society are jointly sponsoring a series of symposia and discussions at their meetings. The first such symposium was held in Albuquerque in 1998, and another one entitled "The Cosmos in the Classroom" was held in Pasadena in July 2000.

Discoveries in astronomy are well covered by the media. For example, staff of the *New York Times* and the *Dallas Morning News*, two of the leading papers in terms of science coverage, each develop on average more than one astronomy story per week. News conferences of the American Astronomical Society are heavily attended, covered by many news media and often held up as a model by other sciences and scientific organizations. Dozens of astronomy columns now run in newspapers and magazines. Many focus on sky phenomena, while others report on recent developments. Perhaps the best known of these is the regular series of science articles published in *Parade*, the national Sunday supplement—a series begun by the late Carl Sagan and now continued by David Levy.

Magazines devoted exclusively to astronomy enjoy wide circulation—nearly 300,000 combined for *Sky and Telescope* and *Astronomy*. Many other national magazines, such as *Popular Science, National Geographic, Discover,* and *Scientific American,* cover astronomy regularly and report that their astronomical stories or issues are among the most popular. It is no coincidence that when *Scientific American* began a new quarterly magazine devoted to single-topic issues, the first was entitled "The Magnificent Cosmos."

Astronomy reaches an extraordinary audience of radio listeners. The program "Earth and Sky" is carried by about 900 radio stations in the United States, and the program is heard about 280 million times each year. "StarDate/Universo" reaches an audience of about 8.7 million listeners weekly. Surveys in Michigan and Florida showed that 51 percent and 36 percent, respectively, of the listeners discussed what they had heard on the "Earth and Sky" program with other adults or children. Eighty percent of the listeners felt the program "expanded their knowledge of science." Gender, ethnicity, and occupational status did not correlate with whether or not a person listened to the series. These statistics show that well-presented astronomy stories have an extremely large and diverse audience.

Astronomical sites are among the most popular science destinations on the Web. The American Astronomical Society has found that news stories carried on Web sites often stimulate stories on affiliated television networks. Web sites offer the additional advantage of coverage in depth since they are not limited in terms of space in the same way as newspapers and television broadcasts. Web sites of the Jet Propulsion Laboratory (JPL) and the Space Telescope Science Institute (STScI) are enormously popular and provide the public with a sense of shared participation in the startling discoveries of planetary probes and the Hubble Space Telescope. For example, the Web provided real-time access for millions to view spectacular events such as the impact of Comet Shoemaker-Levy on Jupiter and the adventures of Pathfinder and Sojourner on Mars. The JPL and the U.S. Geological Survey have developed a planetary photojournal Web site that is accessed by 100,000 users who download 700,000 files every month. These Web sites, as well as those run by the Astronomical Society of the Pacific and the American Astronomical Society, provide resources used by thousands of teachers throughout the nation—and bring the excitement of science from the frontiers of research directly into the classroom.

Public interest in astronomy has fueled a number of successful small businesses. Several hundred million dollars are spent each year by hobbyists, small telescopes users, and travelers journeying to witness astronomical events. The catalog of educational materials in astronomy from the nonprofit Astronomical Society of the Pacific reaches about 300,000 people each year.

ASTRONOMY IN PRECOLLEGE SCIENCE EDUCATION

The national science education standards developed by the National Research Council (NRC, 1996) specify age-appropriate content goals for the teaching of science in grades K-12. However, content goals alone are not enough. Although students may be able to give the correct answers to traditional problems and questions, these correct answers often mask fundamental misconceptions. Largely to address this problem, the national science education standards suggest an emphasis on the teaching of science as inquiry. Engaging students in the active process of inquiry can help them to develop a deeper understanding of both scientific concepts and the nature of science. Through inquiry, students can gain an appreciation of *how* we know *what* we know about science.

Astronomy lends itself extraordinarily well to inquiry-based teaching and allows teachers to take advantage of the natural fascination students have with the field. Many astronomical phenomena can be observed by students directly with no special equipment, and astronomy-based investigations (focusing on topics like light and color, for example; see Figure 4.1) can naturally lead students to explore concepts that inform other scientific fields.

Consequently, astronomers and astronomy educators have invested significantly in developing hands-on activities to support science curricula at all levels. The best of these are collected in *The Universe at Your Fingertips: An Astronomy Activity and Resource Notebook* (edited by A. Fraknoi et al., Astronomical Society of the Pacific, San Francisco, 1995), a resource and activity notebook that is now in use in almost 15,000 schools around the country.

Over the past decade, astronomers also began to work closely with educators to bring data from spacecraft and observatories directly into the classroom and museums (an example is shown in Chapter 5 in Figure 5.2). Programs such as Hands-on Universe (sponsored by Lawrence Berkeley National Laboratory), Hands-on Astrophysics (sponsored by the American Association of Variable Star Observers), Telescopes in Education (sponsored by NASA), and Research-Based Science Education (sponsored by NSF/NOAO) allow students to explore and use newly acquired astronomical data. Simple image analysis tools are now widely available and, when used in connection with images from planetary exploration and telescopic observations, can be powerful tools in engaging the imaginations of students. Programs like these have already led to well-publicized examples of students discovering a supernova and a new Kuiper Belt object. An increasing number of schools are able to connect to the Internet, thereby making access to astronomical data and images widely available.

A number of astronomical organizations and groups have also been working directly with K-12 teachers, providing training, materials, and classroom visits by teams comprising both professional and amateur astronomers (see Figure 5.1 in Chapter 5). By the end of 1999, for

FIGURE 4.1 School children visiting the exhibit Light! Spectra! Action! at the Adler Planetarium and Museum (in Chicago) learn how astronomers use light and spectra to tell the properties of stars. Photographs provided by D. Duncan.

example, Project ASTRO (developed initially by the Astronomical Society of the Pacific) had established about 700 astronomer-teacher partnerships and had reached more than 50,000 students around the country. Through such projects as the AASTRA program sponsored by the American Astronomical Society, the SPICA and ARIES programs at the Harvard-Smithsonian Center for Astrophysics (see Figure 4.2), and the Astronomical Society of the Pacific's Universe in the Classroom workshops, several thousand teachers have learned how to be more effective in conveying astronomy and science to their students. The astronomical community has recognized the value of such efforts and is seeking ways to expand their reach to a larger number of teachers throughout the United States.

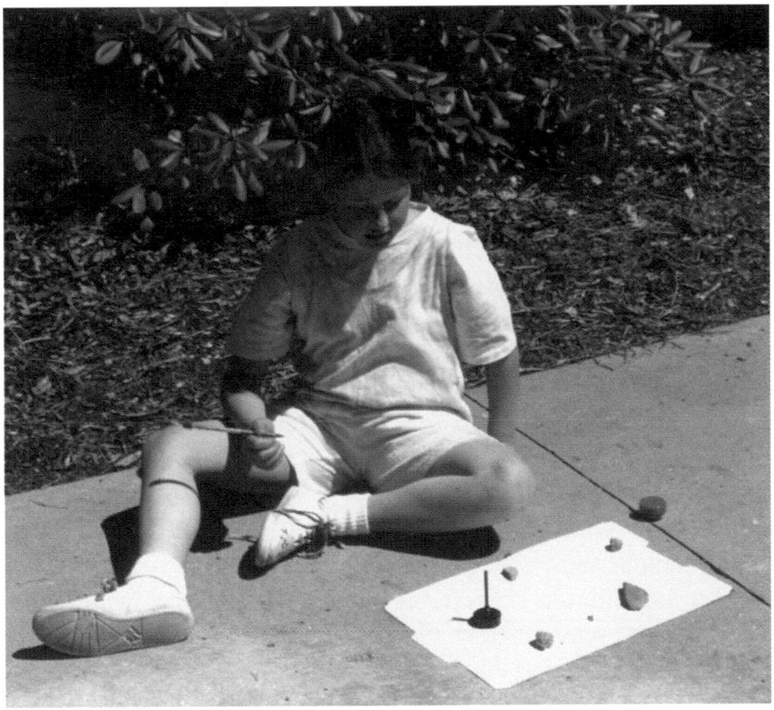

FIGURE 4.2 Elementary school student using a gnomon to follow the motion of the Sun's shadow. This program is a part of the Earth in Motion module of Project ARIES at the Harvard-Smithsonian Center for Astrophysics, Cambridge, Massachusetts. The project is funded by the National Science Foundation, the Smithsonian Astrophysical Observatory, and Harvard University. Photograph courtesy of the Harvard-Smithsonian Center for Astrophysics.

The variety of organized science education outreach efforts built on astronomical themes has been growing rapidly and promises to increase throughout the decade as NASA and the National Science Foundation encourage investigators and teams to add education components to their funded research. A good, frequently updated summary of current national astronomy education projects can be found on the Web site of the Astronomical Society of the Pacific (<www.aspsky.org/education/naep.html>).

Because of the importance of linking the public investment in research to advancing public science education goals, the astronomical community has worked hard to identify areas where successes have been achieved, efforts that are highly leveraged, and ways that those gains can be propagated. Recommendations aimed at better coordinating these efforts in the new decade are described in Chapter 5.

THE PRACTICAL CONTRIBUTIONS OF ASTRONOMY TO SOCIETY

Federal support of curiosity-driven scientific research has historically led to a broad range of contributions to technological advances with long-term benefits to society. Indeed, national investment in curiosity-driven scientific research is widely viewed as an essential element of U.S. economic strength and competitiveness. Despite its focus on the extraterrestrial, astronomy has made important contributions on Earth as well. In large measure, these contributions derive from the need to measure precise positions, luminosities, and structural details in faint and distant cosmic sources, to measure time with exquisite precision, and to analyze large statistical samples of objects spanning a wide range of physical, chemical, and evolutionary conditions. All these activities have led to numerous benefits to society that are discussed in more detail below. In some areas, astronomers have pioneered the technology, while in others they have worked symbiotically with industry and the defense sector in developing and perfecting the appropriate technologies.

ANTENNAS, MIRRORS, AND TELESCOPES

Large mirrors or antennas that focus and image light, infrared radiation, or radio waves are used not only by astronomers but also by, for example, the communications industry, the military (e.g., in surveillance), and the scientists who use telescopes that look down from space to study Earth's ecosystem and resources. In order to produce a sharp image, either large-diameter mirrors or antennas are required, or the radiation must be collected on widely spaced individual mirrors or antennas and then combined—a technique called interferometry.

Besides size, another key to a high-quality image is producing a very accurately shaped mirror or antenna. Astronomers have made major contributions to mirror and antenna technology. Examples include developing mirror materials (lightweight materials in particular), mirror designs, precision shaping and metrology (shape testing), procedures for correcting the effects of bending under the force of gravity, technologies to correct for the blurring effect of the atmosphere (e.g., a technology called adaptive optics), interferometry, and the technology for steering the beams and efficiently collecting the radiation in large radio telescopes. Besides the obvious applications noted above, there are additional spinoffs. One notable example is in the area of adaptive optics. Techniques developed by astronomers for adaptive optics are being refined to produce ophthalmic instruments that can image the retina of an eye and measure an individual's eye aberrations in unprecedented detail. The potential exists for low-cost diagnosis of eye disease, as well as for specification of parameters for either contact lenses that will provide "supernormal vision" or corrective eye surgery.

Adaptive optics techniques and techniques to manufacture and figure ultralightweight, ultrahigh-precision mirrors are examples of synergy between investments in defense-related technology and in astronomy. The rapid growth of adaptive optics over the past decade owes much to the declassification of techniques developed in the service of national security interests. Mirrors for the Hubble Space Telescope are a direct descendent of efforts in service of surveillance during the 1970s and 1980s, while today, NASA and the National Reconnaissance Office are partners in efforts to develop next-generation, large space-based mirrors.

SENSORS, DETECTORS, AND AMPLIFIERS

Perhaps the biggest technology spinoff contributed by astronomy has been the development or improvement of devices that convert light and other forms of radiation into images. Historically, astronomy pushed the development of photographic film to greater sensitivities and resolution. However, film has now been largely replaced by electronic sensors, detectors, and amplifiers—devices that enable accurate digitized measurements of brightness over a wide range of wavelengths. In this section, astronomy's contributions to signal detection are discussed by frequency band, starting with the high-frequency x-ray band and moving to ever lower frequencies: ultraviolet/optical, infrared, and radio.

X rays partially penetrate opaque objects and can thus be used to image their "insides." One prominent example is provided by the luggage scanners used as security devices in airports. The most common version of this device is a spinoff from space x-ray astronomy, where the requirement to observe weak cosmic signals resulted in the development of high-sensitivity x-ray detectors. Application of these detectors to luggage scanners enabled the use of low x-ray dosages to obtain good images, thus enhancing their safety for operators and passengers alike. X-ray astronomy detectors, with their sensitivity to single photons and to low-energy x rays, are also ideally suited for fundamental biomedical research, for cancer and AIDS research, and for drug and vaccine development. These sensitive detectors have led to a plethora of x-ray medical imaging devices, including those used to search for breast cancer, osteoporosis, heart disease (the thallium stress test), and dental problems. The last is a new development that uses x-ray charge-coupled devices (CCDs; miniature electronic detectors) to replace dental x-ray film, a change that will reduce exposure to x rays. Another exciting development is the x-ray microscope. A microscope is, in effect, a miniature telescope. X-ray astronomy has led to the development of the Lixiscope, a portable x-ray microscope to be used to image small objects and fine detail, with applications in energy research and biomedical research. It is widely used in neonatology, out-patient surgery, diagnosis of sports injuries, and Third World clinics. The Lixiscope is NASA's second largest source of royalties. In a somewhat different technique called x-ray diffraction, a "super-microscope" is achieved that is capable of studying tiny molecular structures. This technique utilizes the interference of the x rays with each other after they scatter off a sample surface. X rays are preferred because they resolve molecular structure. Astronomical advances in detector sensitivity and focused beam optics have allowed the development of systems with much shorter exposure times, and have allowed researchers to use smaller samples, avoid damage to samples, and speed up their data runs. Biomedical and pharmaceutical researchers have used these systems for basic research on viruses, proteins, vaccines, and drugs, as well as for cancer, AIDS, and immunology research.

At ultraviolet (UV) and optical frequencies astronomers have pushed the development of more sensitive CCDs and of large arrays of CCDs.

Cooled silicon CCD arrays developed for optical astronomy now dominate in a multitude of industrial imaging applications. The basic performance of these detectors has been improved by a thinning process developed by astronomers. CCD manufacturers have adopted this technique for use on Earth satellites (e.g., to watch for lightning strikes in the atmosphere) and in surveillance applications. In the UV, CCD development undertaken for a Hubble Space Telescope instrument was later incorporated in a stereotactic breast biopsy machine, which detects tumor positions accurately enough to steer the biopsy probe, thereby reducing the need for surgery and cutting costs by 75 percent (see the Scientific Imaging Technologies Web site at <www.site-inc.com/news-breastcancer.htm>). In addition, UV detectors developed for the Hubble Space Telescope are being considered as a key element in a helicopter-based system aimed at rapid detection of power-line failures in remote areas.

Objects on Earth radiate most of their energy at infrared (IR) frequencies. In addition, infrared radiation can in some cases be more penetrating than visible light, thus rendering it useful for looking "inside" objects, in analogy to x rays. For both of these reasons, the development and/or improvement of sensitive IR detectors, large-format arrays, and IR techniques by infrared astronomers has had significant benefit to society. In this area, there has been a symbiotic relationship with the Department of Defense, which has invested large amounts of money in IR detector development for defense applications. Improvements made by astronomers have contributed to the final versions of the detectors used in the Strategic Defense Initiative and for night-vision devices. In the industrial sector, IR detector arrays developed by astronomers are being used in the semiconductor industry in IR microscopes that examine computer chips for flaws. In the medical sector, IR detectors and spectroscopes are being used to diagnose cervical cancer and genetic diseases and to image malignant tumors and vascular anomalies.

Not only radio and television, but also all satellite and much telephone communication is accomplished with radio waves. Radio astronomers have provided the impetus to many technical advances that have improved the stability, widened the bandwidth, and reduced the noise and interference of radio communications: low-noise maser, parametric, and other transistor amplifiers that have had wide application in the communications industry. Astronomers have perfected high-radio-frequency systems that have found application in devices to detect concealed weapons, to see through fog and adverse weather for aircraft landing systems, and to image human tissue (e.g., in mammograms).

SPECTROMETERS AND DEVICES TO FOCUS RADIATION

Astronomers have driven the development of ever more precise instruments, called spectrometers, that separate and analyze the different frequencies present in a beam of radiation. In addition, they have perfected precision techniques to focus radiation into spots too small to be visible. These developments have been highly beneficial to the industrial, defense, and medical sectors of the economy.

NASA supported the development of a novel x-ray spectrometer, the microcalorimeter, for x-ray astronomy, but this new device can also be used to analyze the chemical elements in a small sample. Applications include materials science research, rapid trace-element analysis for the semiconductor industry (semiconductor wafer testing), and biomedical research, which requires low doses for biological samples. X-ray spectrometers developed in part in response to the needs of astronomy are also used in x-ray laser materials science and in fusion energy research, as well as in the nuclear nonproliferation program. UV spectrometers are used in laboratory analysis equipment. IR spectrometers remotely analyze the composition of the atmosphere. Spaceborne and ground-based radio spectrometers remotely monitor temperature, winds, humidity, and chemical composition in the atmosphere with applications to weather prediction, global warming, and pollution monitoring. The depletion of ozone has been monitored with astronomical radio telescopes equipped with radio spectrometers. Spaceborne radio spectrometers also sense ground-level quantities such as soil moisture, vegetation cover, ocean height and sensitivity, oil spills, snow cover, and iceberg hazards. Essential components of all these spectrometers have been invented or perfected by the astronomical community.

Efforts in UV and x-ray astronomy pioneered the development of technologies crucial for UV and x-ray lithography, a process by which fine beams of radiation etch lines in a material. Very fine line widths are needed by the semiconductor and microchip manufacturing sector to make advanced computer chips, transistors, and other microelectronic devices. In the medical sector, astronomical technology invented to focus x rays is being put to use in precision deposition of x-ray radiation to destroy cancerous tumors.

IMAGE RECONSTRUCTION

Astronomers are bedeviled by faint and blurred images that are often swamped by large amounts of noise or static. An analogous problem would be faint TV reception, superimposed on the static produced by a hair dryer operating nearby. Consequently, astronomers have been at the forefront of efforts to improve and sharpen images, to reduce extraneous noise, and to extract the maximum information from the radiation received. One example of this effort is a system of image analysis tools and computer applications programs developed by astronomers at the National Optical Astronomy Observatories: IRAF, the Image Reduction and Analysis Facility. IRAF has been used not only by thousands of astronomers worldwide, but also by researchers outside astronomy engaged in underwater imaging, mapping of the aerosols in the atmosphere, medical imaging for detection of breast cancer, decoding of human genetic material (in connection with the Human Genome Project), numerous defense-related applications, visualization of the images from electron microscopes, and many other applications. AIPS, the Astronomical Image Processing System developed at the National Radio Astronomy Observatory, is another software package for manipulation of multidimensional images that is used routinely in nonastronomical image analysis applications. Astronomers have also contributed to the advancement of tomography, which enables construction of three-dimensional images out of a series of two-dimensional pictures. Tomographic imaging is used widely in both medical x-ray imaging and industrial applications. The image reconstruction work of R. Bracewell, a pioneering radio astronomer, is widely cited by the medical imaging community. Techniques pioneered by astronomers, such as "wavelet smoothing" and "maximum entropy," have been used for pattern recognition in areas like mammography and to sharpen images for police work.

PRECISION TIMING AND POSITION MEASUREMENTS

Interferometry is the main technique used by astronomers to measure with ultrahigh precision the position in the sky of astronomical objects. Interferometers employ two or more telescopes located some distance apart that precisely measure the time difference in the arrival of radiation from a source. To do this properly requires extremely accurate clocks, since the time differences are extremely short. Astronomers played a significant role in refining the hydrogen maser clock, which is now widely used for space communications and in the defense sector. The interferometric timing technique to locate radiation sources has had widespread application, including finding noise sources (such as faulty transmitters that interfere with communications satellites), locating cellular phones to track locations of 911 calls, measuring the tiny shifts of Earth's crust before and after earthquakes, and precisely locating people and vehicles using the Global Positioning System precision surveying network.

DATA ANALYSIS AND NUMERICAL COMPUTATION

Astrophysics has been a major driver of supercomputer architecture and computational science for nearly 50 years. Computations of stellar evolution by the pioneering astronomer Martin Schwarzschild occupied nearly half of the time of one of the first computers (MANIAC). Computers are severely challenged by the gigabytes of data streaming in daily from modern astronomical sensors and large sky surveys, and by the large computational speeds required for both simulations and database searches. These requirements are stimulating the development of large computers and innovative hardware components. Beowulf computers, which provide simple commodity supercomputing, were developed by astronomers to enable sophisticated numerical simulations. The idea of designing special-purpose hardware for a specific task has also flourished in astronomy. Two examples of such hardware are the GRAPE computer chips for doing large-scale gravitational N-body simulations (details are available at <grape.c.u-tokyo.ac.jp/grape/>), and the Digital Orrery for calculating the motions of the bodies in our solar system (now retired at the Smithsonian Institution in Washington, D.C.). The Gordon Bell Prize—a prestigious award for significant achievement in the application of supercomputers to scientific and engineering problems—was won by astronomers in 1992, 1995, 1996, 1997, and 1998. FORTH, a high-performance computer programming language and operating system, was developed at the National Radio Astronomy Observatory and has been used in hand-held computers carried by Federal Express delivery agents and by automotive engine analyzers in service stations, in environmental control systems in airports, and by Eastman Kodak in quality control for film manufacturing.

Many software developments were also either created by astronomers or received much of their impetus for improvement from them. Fast Fourier transforms and other image-processing techniques were greatly improved by radio astronomers and later by optical astronomers. Some of the more popular grid-based computational fluid dynamics techniques that are used in applications such as weather prediction were either created or improved by astronomers. Another particle-based hydrodynamic technique, smoothed particle hydrodynamics, was both invented and improved by astronomers and has found uses outside astronomy, for example in modeling ballistic impacts. Magnetohydrodynamic codes and numerical simulations of plasmas developed by astronomers contribute to design efforts aimed at harnessing fusion power. Digital correlation techniques for spectral analysis of broadband signals have been adapted for use in remote sensing, oceanography, and oil exploration. IDL, a commonly used graphical package, originated as visualization software for the Mariner Mars 7 and 9 space probes. "Numerical Recipes,"[1] a collection of numerical algorithms that is now widely used throughout science, started as an astronomy course on scientific computing. To handle the large databases being produced by astronomical surveys, several groups are collaborating with computer scientists to push forward the frontiers of database mining. Inexpensive and error-free methods of archival mass data storage have been invented by astronomers. Such developments will obviously have far-reaching applications. Finally, astronomy serves as a prolific and productive training ground for many computational scientists.

EARTH'S ENVIRONMENT AND PLANETARY SURVIVAL

Astronomical studies are essential to understanding the evolution of Earth's atmosphere and the factors that drive climate changes. Geological evidence suggests that in past millennia, Earth's climate—as well as the atmosphere and oceans that control it—was remarkably different. It is now certain that the astronomical environment, including changes in the Sun's brightness, the influx of cosmic rays, variations in Earth's orbit, and the influx of zodiacal dust, is an important driver of major long-term climatic changes, such as the ice ages, as well as some smaller and more rapid changes. Together, astronomical and geological observations provide the framework for understanding the response of the biosphere to external change, which is an essential precursor to comprehending and predicting the relative importance of changes that may be wrought by modern industrial activity.

Deepening our understanding of the factors that control climatic conditions on Earth will depend critically on continued careful observation of the Sun itself, of variations in luminosity among other similar stars spanning a wide range of ages, and of the distribution of dust in the solar system. Observations of atmospheres surrounding other planets combined with numerical modeling also promise important insight into the complex interactions that drive climate changes.

In addition to terrestrial climate, understanding and monitoring solar "climate"—the ebb and flow of energetic particles arising in solar flares and the solar wind, and the ultraviolet output of the Sun—are also essential. The effects of energetic particles on radio communications can be dramatic, and variations in solar ultraviolet radiation can play a major role in affecting the concentration of critical trace atmospheric constituents such as ozone.

Study of the history of collisions of asteroids and comets with Earth provide the framework for understanding *cataclysmic* climate changes over geological time scales. While far rarer now than during the first billion years in the solar system's history, collisions of comets and asteroids with planets still take place. On Earth, such collisions can produce dramatic environmental events, from giant tidal waves to Earth-girdling dust clouds that can alter climate for centuries and in some cases lead to mass extinctions of species. Astronomers now have the tools to detect comets and Earth-crossing asteroids of size sufficient to threaten human civilization and to assess the threat of a collision. In Chapter 3, the report discusses the potential role of the Large-aperture Synoptic Survey Telescope in providing a census of Earth-crossing asteroids.

CONNECTIONS BETWEEN ASTRONOMY AND OTHER DISCIPLINES

Building an understanding of the universe and the evolution of its constituents requires tools and insights from many other disciplines, including physics, chemistry, optical science, electrical engineering, computer science, and biology. In turn, astronomical discoveries as well as the need to develop new instruments and computational techniques often provide strong impetus for developments in other disciplines.

INTERACTIONS WITH PHYSICS

The largest set of interactions of astronomy with other fields currently involves modern theoretical physics, whose major goal is to understand the basic constituents of matter and the forces between these constituents. The classic example is the 300-year-old problem of the stability of the solar system. Astronomers working on this and other problems in astrophysical dynamics have provided much of the inspiration for and many of the tools used in modern nonlinear dynamics and chaos theory, which are now routinely applied to subjects as diverse as evolutionary biology and the stock market. In turn, the concepts of nonlinear dynamics have been successfully used by astronomers in the past decade to determine the dynamical stability of planetary systems, including our own.

One of the greatest intellectual advances of the past millennium was the recognition that all of the laws of nature work the same way in both the laboratory and the cosmos. However, the universe provides higher energies, temperatures, and densities; stronger magnetic and gravitational fields; and much longer observing times than we can hope to reproduce on Earth. Important findings include (1) the discovery that nuclear fusion provides the energy source for the Sun and other stars, (2) evidence from the Sun and cosmic rays that the three sorts of neutrinos can interchange their identities, (3) observations of changes in the orbits of binary pulsars (neutron stars) that provide dramatic confirmation of gravitational radiation, a crucial prediction of Einstein's theory of relativity, (4) very tight limits on the extent to which the strengths and other properties of the forces could have changed over billions of years, and (5) the recognition that most of the matter in the universe resides in some mysterious unseen form ("dark matter"), perhaps a new kind of elementary particle, and the recent evidence that a novel form of "dark energy" dominates the dynamics of the cosmic expansion.

The committee agreed that astronomers and astrophysicists can reasonably anticipate a number of future interactions with physics:

- In the realm of very high energies, high energy densities, and high pressures common in astronomical objects, where advances can illuminate areas such as nuclear physics, high-energy physics, and new states of matter;
- In investigations of plasma dynamics, energetic fluid behavior, magnetic interaction with matter, turbulence, and chaos—phenomena whose complex dynamics represent one of the major scientific and engineering challenges today, and one where astronomical examples and theory are being studied intensively; and
- In "astronomical laboratories" that extend the reach of terrestrial

laboratories by providing opportunities to test the known laws of physics in extreme environments (e.g., the strong gravitational fields near the event horizons of black holes or the high density of neutron stars) or to search for new physics such as new particles, new forces, and the unification of forces (e.g., solar neutrinos, the early universe).

ASTRONOMY AND THE COMPUTATIONAL SCIENCES

The astronomical community stands poised to take advantage of the continuing breathtaking advances (factor-of-2 increases each 18 months) in computational speed, storage media, and detector technology in two ways: first, by carrying out new-generation surveys spanning a wide range of wavelengths and optimized to exploit these advances fully (see Chapters 1 and 3); and second, by developing the software tools to enable discovery of new patterns in the multi-terabyte (10^{12}) and, later, petabyte (10^{15}) databases that represent their legacies (Chapter 3). In combination, new-generation surveys and software tools can provide the basis for enabling science of a qualitatively different nature. Whereas in the past astronomical experiments were constrained by the need to carefully select small samples, often strongly guided by a priori assumptions, astronomers can now plan far more objective approaches based on deep images of wide areas of the sky spanning a range of wavelengths, or on spectra of millions of stars and galaxies.

Here, the committee noted the potential synergy of these efforts with computer science and other scientific disciplines facing the need for

- Simultaneous, rapid querying of individual terabyte archives by thousands of researchers located at remote sites throughout the world;
- Complex querying of multiple catalogs and image databases, including efficient correlation of catalog and image information from these archives aimed at discovery of complex patterns or rare phenomena through advanced visualization and sophisticated statistical tools—to discover rare galaxies that "look like this, but not that."

Developing efficient archiving protocols, querying methods, and data mining and visualization tools will enable astronomers to fully exploit the rapid advances in computer and detector technology. Forging partnerships among scientists who are working on these fundamental issues will enable linkage of the efforts of multiple scientific communities with the dynamism of commercially driven efforts to address these problems. For example, the kinds of tools needed by astronomers to query huge image databases are closely related to those needed by physicians searching for commonalities and anomalies in medical images, and by agricultural analysts studying crop yields by analyzing satellite images.

POTENTIAL INTERACTIONS WITH THE BIOLOGICAL SCIENCES: ASTROBIOLOGY

Astronomy and space science have reached a stage where astronomers can make important contributions to answering fundamental questions related to the origin and distribution of life in the universe. For the first time, astronomers are able to trace, both observationally and theoretically, the birth of planetary systems around other stars. Researchers are also able to determine the key events in the history of Earth and other planets in our solar system, including the sources of water and other volatiles, and of the organic chemicals that are the building blocks of life. In the past few years, astronomers have discovered around other stars more than five times as many planets as the nine planets comprised by our solar system. Powerful new instruments will soon permit us to survey for Earth-like planets around Sun-like stars.

Within a decade astronomers may be able to search for the spectroscopic signatures of biogenic gases, which provide evidence for life on such extrasolar planets. But researchers can recognize the signature of life elsewhere only by understanding better the history of life on Earth over the past 4 billion years and exploring more deeply the possibility that life has also had an independent history on Mars or other planets and moons in our solar system. This study is an essential part of the new synergy between astronomy, planetary science, and biology—what has been called astrobiology. This nascent activity aspires to encourage collaborations across these disciplines in order to address questions that compel the imaginations of scientists and citizens alike: What is the origin and evolution of life? Is there life elsewhere in the universe? What is the future of life on Earth and in space?

Answering these questions will require the combined efforts of astronomers, planetary scientists, and biologists, as well as investments in new facilities and instruments. Astrobiology has the potential to draw together investigators from disciplines that in the past have shared little except a common interest in understanding the natural world. Astronomers can contribute to this effort by determining the conditions that lead to the formation of habitable planets, by finding planets and satellites that could be habitable, and by searching for evidence for life through remote observation. It is difficult to predict the potential for advances at this new interface between the physical and life sciences—other than to note the extraordinary potential of bringing diverse scientific cultures together at the right moment in time.

Perhaps of equal importance is the extraordinary public interest generated by attempts to understand our origins and the ubiquity of life in the universe. Like astronomy, astrobiology has the potential to link the seemingly abstract world of research at the frontiers of knowledge to questions that have excited the human imagination since people first gazed at the heavens.

NOTE

1. "Numerical Recipes" refers to copyrighted software published in the series *Numerical Recipes: The Art of Scientific Computing*, available as books from Cambridge University Press and also in electronic form at <www.nr.com>.

5

The Role of Astronomy Education

INTRODUCTION

A challenge facing our country in the new century is to maintain the integrity and vitality of the scientific and technological leadership that the United States has enjoyed in the past 100 years. An essential component of a healthy scientific enterprise is a scientifically literate and well-educated public, and professional scientists have a vital role to play in achieving a world-class system of science education. In the past decade, the growing consensus about the need for rejuvenation of science and mathematics education in our nation's schools led professional scientists to join with educators in taking a careful look at how science is taught and how teachers are trained to teach science. One result of efforts made by the National Science Foundation (NSF) and the Department of Education, along with the National Academy of Sciences and the American Association for the Advancement of Science, has been the development of national benchmarks and standards for science education.

Despite its comparatively small size, the astronomical community has the potential to add significantly to the continuing effort to strengthen science education and improve public science literacy. Astronomical concepts and images have universal appeal, inspiring wonder and resonating uniquely with human questions about our nature and our place in the universe. This widespread interest in astronomy can be tapped not only to increase knowledge and understanding on the part of students and the public alike, but also to illuminate the nature of science, as well as its power and limitations in shaping our future. Moreover, the interdisciplinary nature of astronomy and its natural links with technology and instrumentation position the field to contribute significantly to building a strong technical work force for the 21st century.

Astronomers are keenly aware of the generous support provided by the public through the federal science agencies, and they understandably wish to contribute to the society that supports their research activities. Education and public outreach provide means to do so. The astronomical community's mobilization on the education front has begun with major educational initiatives undertaken by NASA's Office of Space Science, the Astronomical Society of the Pacific, and the American Astronomical Society, and astronomers are participating actively in educational opportunities offered by the NSF. For example, Project ASTRO of the Astronomical Society of the Pacific brings professional astronomers and teachers together in workshops to learn hands-on, inquiry-based techniques for teaching science (Figure 5.1). Project ASTRO was begun with NSF support.

Full realization of the astronomical community's potential to advance science education requires a clearly envisioned mission and goals, coupled with a focused and coordinated set of high-leverage investments designed to maximize astronomers' contributions to all levels of science education. To this end, the committee describes the educational mission for the astronomical community as the pursuit of four broad goals:

1. To disseminate astronomical discoveries widely, and thus bring the excitement inherent in science to the American public.

2. To use the excitement that astronomy engenders to increase public understanding of science and scientific methods and to make clear that science is a pathway to discovery, not just a collection of facts. This must be done at both the K-12 level and the undergraduate level.

3. To capitalize on the close involvement of astronomy with technology and instrumentation to contribute to training the technical work force.

FIGURE 5.1 Teachers learn about the properties of light and color at a Project ASTRO workshop conducted by the Astronomical Society of the Pacific. Partnered astronomers and teachers are trained together at 2-day workshops and then develop an individualized educational program for the teacher's class. Photograph courtesy of the Astronomical Society of the Pacific, R. Havlen.

4. To prepare future generations of professionals who will sustain U.S. preeminence in astronomy and will contribute to a scientifically literate nation.

Laying out strategies to achieve these educational goals and then describing existing programs and future directions, this chapter follows closely the report prepared by the Panel on Astronomy Education and Policy.

STRATEGIES TO ACHIEVE THE FOUR EDUCATIONAL GOALS

COMMUNICATE DISCOVERIES AND EXCITEMENT OF SCIENCE

Astronomical discoveries captivate the human imagination by connecting to deep and long-standing questions about our origins and the nature of the universe in which we live. Awareness of the vastness of the universe, the extraordinary forms that other worlds can assume, and the place of our home planet in space and time inspires wonder in people of all ages. With the exception of medicine, no other scientific discipline has seen its new accomplishments featured so often on the front pages of national newspapers and the covers of national magazines. Moreover, astronomy has become the focus of an array of publications dedicated to amateurs and students. Astronomy's natural appeal and popularity in fact underscore professional astronomers' important responsibilities in seeking to ensure that the public is kept abreast of the latest advances and can appreciate their relevance within the larger context of natural science.

At the heart of enhancing public awareness of and appreciation for science is effective communication with the general public about discoveries made in the research community. Responsibilities for broad dissemination of new knowledge in astronomy are shared by the agencies that support research and by scientific and academic institutions, professional organizations, and individual astronomers. Cogent accounting of the benefits of the public's investment in major NASA space-based and NSF ground-based astronomy facilities is particularly important (Chapter 4 briefly discusses many of these benefits). Tremendous public interest in the scientific accomplishments of NASA's space missions has been sustained and complemented by imaginatively publicized and broadly distributed news of these achievements. The committee believes that enhanced public awareness of the equally notable achievements of NSF-funded science from optical, infrared, and radio astronomy ground-based facilities is essential. The importance of broad visibility is clear in an era when all avenues of federal spending are scrutinized by a fiscally responsible Congress.

- **The committee recommends that the National Science Foundation invest in advancing broad understanding and appreciation of science by improving public recognition of the achievements of NSF-funded science and facilities, with initial emphasis on subjects with wide public appeal, such as astronomy.**

Better communication with the public would require that the NSF establish a direct interface with the media by appointing several dedicated press officers who would stay in close communication with NSF program officers and individual scientists and would be assured of adequate technical support. In addition, NSF should strengthen its presence on the Internet by developing informative and stimulating Web pages addressed to the public and designed to dramatize the scientific achievements sponsored by the NSF. The effort to communicate more effectively should also include increased investments in state-of-the-art displays at centers and facilities supported by the NSF. Sophisticated Web pages dedicated to informing the public of the scientific discoveries made at these facilities must be maintained, and outreach into the local communities should be supported.

Centers of informal science education offer another important means of communicating science to the public. There is a need to promote and strengthen communication between professional scientists and the planetariums, museums, and science centers visited by people of all interests and ages (see Figure 4.1 in Chapter 4). In the past decade several of the nation's leading planetariums took major steps to improve their interaction with professional astronomers and hired a small staff of research scientists. Most museums and planetariums, however, currently lack the scientific expertise to inform their visitors succinctly about advances in modern astronomy. At the same time, most scientists find it difficult to tune their knowledge to the diverse backgrounds and range of experience represented by the many visitors to such centers. Cooperative efforts are thus essential to realize the maximum potential of science museums. The committee notes that some important steps in this direction have been taken by the Informal Science Education program at the NSF and by NASA's Office of Space Science (OSS) Educational Ecosystem initiative, as well as by some museums.

The committee recommends stimulating and enhancing interactions between science education institutions and the research community. Expanded federal support would enable more robust programs to be developed. In particular, the American Astronomical Society (AAS) and the Astronomical Society of the Pacific (ASP) could take a lead role in sponsoring workshops for museum staff led by scientists and educators, helping the staff to develop displays and to disseminate informative images and interactive software to the nation's science centers.

"Project ASTRO is a national program to help improve the teaching of astronomy and physical science in general in 4th through 9th grade classrooms (and youth groups). The main focus of the project is on hands-on, inquiry-based activities that put students in the position of acting like scientists as they come to understand more about the universe (and science in general). Professional or amateur astronomers are linked with local teachers or youth group leaders, and 'adopt' a classroom or community group, visiting 4 to 10 times per year."

—Excerpted from the Project ASTRO Web site at <www.aspsky.org/project_astro.html>.

Professional organizations, academic departments, and individual astronomers all need to be active in communicating scientific discoveries to the public. The AAS and the ASP have done an outstanding job for the astronomical community in this arena, through press releases, publications, and workshops for teachers. There remains, however, a need for both academic departments and individual astronomers to strengthen their commitment to communicating the excitement inherent in astronomy to their local communities. The committee urges all astronomy departments to invest in this important enterprise at a level commensurate with their size. The AAS can play a coordinating role, sharing examples of activities led by departments and individual astronomers that have proved particularly effective in sparking excitement about science in local communities. The AAS already maintains an astronomy education database and is currently investigating the possibility of developing a journal for astronomy education. The committee also urges all graduate departments involved in astronomy and astrophysics to encourage their students to gain experience in public outreach, to ensure that future generations of astronomers develop a sense of responsibility for contributing to the public's understanding of science.

EXPAND OUTREACH TO K-12 STUDENTS

The nation is undergoing a crisis in K-12 science education, but astronomy, although it holds great fascination for young people, plays a comparatively small role in the formal K-12 curriculum. Indeed, in recent decades its role has diminished as a result of curricular reforms. To reverse this trend, the astronomical community must take a proactive role to ensure that the educational advantages of using astronomy as a gateway into science are not abandoned by the K-12 community. While the astronomical community is eager to play a role in K-12 outreach, it still has much to learn about identifying the most effective, highly leveraged ways for scientists to contribute.

- **The committee recommends expanding and improving the engagement of astronomers in outreach to the K-12 community by ensuring (1) appropriate incentives for their involvement, (2) training and coordination for effective and high-leverage impact, and (3) recognition of the value of this work by the scientific community.**

To date, some of the best efforts in K-12 outreach by astronomers have been in developing age-appropriate, astronomy-based educational materials in partnership with educators. Some of these efforts have been led by dedicated individuals, and others are coupled directly to NASA space missions and facilities (see Figure 5.2). Development of curricular materials that will convey a good understanding of basic scientific concepts is an important first step, and there is certainly a need for additional effort in this direction. Far more difficult tasks are (1) finding the best way to ensure that excellent materials are widely disseminated and adopted, and that teachers know how to use them, and (2) educating astronomers who are eager to reach out to their local communities so they can make the most effective use of their time. Both include the goal of preventing astronomers and educators from unnecessarily duplicating past efforts and thus "reinventing the wheel."

The first of these formidable tasks is being tackled by NASA's OSS Educational Ecosystem initiative, one of the most ambitious educational programs involving the astronomical community. The goals of this program are excellent—to foster a wide variety of highly leveraged education and public outreach activities and to disseminate them in school systems across the country. Because this program is still ramping up, its success has not yet been demonstrated. To ensure maximum

FIGURE 5.2 New York City junior high and high school students assembled in the Hall of Meteorites of the American Museum of Natural History for a live Public Broadcasting System program during the Mars Pathfinder landing in the summer of 1997. A network of museums across the country were connected live to the Jet Propulsion Laboratory for the event. Photograph courtesy of the American Museum of Natural History Library.

benefit for all communities, programs of this magnitude must be subjected to careful scrutiny by both scientists and educators.

- **The committee recommends an external review of NASA's Office of Space Science Educational Ecosystem program early in the decade by both educators and astronomers, using assessment standards agreed to by both groups.**

Professional societies can play an important coordinating role in assisting astronomers who wish to contribute to education and outreach.

- **The committee recommends that the American Astronomical Society, in cooperation with the Astronomical Society of the Pacific, play a lead role in aggressively searching for**

exceptionally effective K-12 outreach programs, and then work to see them adopted widely by members of the astronomical community interested in pursuing outreach activities.

Most funding opportunities for involvement of professional scientists in K-12 outreach programs are limited to seeding new initiatives. At the same time, a growing number of programs are recognized as highly successful, but no opportunities exist for extending them beyond the initial funding period. The insistence on innovation as a major criterion for funding educational projects is shortsighted, and funding agencies should take the lead in coordinating the efforts of federal, state, and local agencies to ensure the preservation of successful programs.

- **The committee recommends that federal agencies explore mechanisms to leverage federal funds to provide long-term support for successful outreach programs in science education.**

IMPROVE SCIENCE LITERACY FOR UNDERGRADUATES

The considerable public interest in astronomy provides an invaluable opportunity to go beyond simply informing and exciting the public about the latest scientific discoveries. Imaginative use of astronomical imagery and phenomena can provide a gateway to increase scientific *understanding*, by clarifying how nature behaves and how the scientific method leads us to develop models of this behavior and then subject these models to rigorous tests. While well-designed public Web sites can contribute to progress in this arena, conveying a true understanding of science requires more formal educational settings. The AAS's Education Office and the ASP have undertaken ambitious programs, including training new faculty and holding periodic workshops for college teachers, to engage their members in efforts to improve formal science education both at the college level and through outreach to the K-12 community. These programs require further support.

At the college level, where many professional astronomers are actively engaged in education, astronomy is one of the most popular science electives, with more than 200,000 students per year enrolled in introductory classes. For many college students, astronomy is their only encounter with a natural science (Figure 5.3). And, given its interdisciplinary links with other fields including physics, mathematics, and geology,

FIGURE 5.3 Two undergraduate students on the Williams College expedition to Rimnicu Vilcea, Romania, for the August 11, 1999, total solar eclipse. They are adjusting a 14-inch telescope for an experiment to detect coronal oscillations during the phase of totality. The expedition was supported in part by grants from the National Science Foundation, NASA, and the National Geographic Society's Committee for Research and Exploration. Photograph courtesy of J. Pasachoff (Williams College).

astronomy is a particularly appropriate vehicle for teaching science to a wide audience. By many measures, astronomy plays a very positive role in general science education at the college level and clearly attracts many students. However, there is a growing awareness that the traditional lecture format coupled with a broad survey of astronomical topics has limited pedagogical success. A national dialog focused on the effectiveness of this popular survey course would allow astronomers from a wide range of institutions to evaluate whether their current practices are the best way to teach science to general college audiences.

- **The committee recommends that the American Astronomical Society and the Astronomical Society of the Pacific cooperatively conduct a national examination of effective ways to improve the understanding of fundamental scientific concepts delivered in popular introductory astronomy classes.**

One of the most important college audiences that can be reached with introductory astronomy classes is America's future elementary and high school teachers. These are the people who will soon be teaching science to the nation's children, and their experience with science in college is of paramount concern. Indeed, it is only by working closely with this audience that we can expect to achieve long-term systemic reform in K-12 science education. The NSF's Education and Human Resources Directorate has recognized this need and sponsors a program designed to involve scientists and education departments collaboratively in improving teacher preparation in science. The subject of astronomy is particularly well suited to form the basis for exposing preservice teachers to interactive inquiry-based teaching. When introductory astronomy courses are designed, special consideration should be given to the particular kinds of science teaching skills needed by the nation's future teachers. In some universities, partnerships between departments of education and departments of astronomy have begun to explore how scientists can become more closely involved in the training of the nation's teachers; the committee applauds these efforts.

- **The committee recommends that more universities with both astronomy and education departments establish pilot partnerships to bring scientists, educators, and experienced teachers together to design exemplary astronomy-based science courses for teachers in training (preservice) with the goal of contributing to long-term systemic reform in K-12 science education.**

A more limited, but still important, college audience is students in schools of journalism. Collaborative efforts between astronomers and journalism programs would also be an investment in the future. The goal of courses directed toward this audience might differ from the goal of those for future teachers and could be considered in the context of a national dialog on introductory undergraduate courses.

Both in and out of the college classroom, extensive and effective use of computing technology can enhance understanding of basic science.

The simple beauty of many astronomical images makes them particularly effective in attracting attention to an astronomical discovery, which can then be used to clarify basic scientific principles. Computers offer a powerful laboratory instrument for this instruction, given their ability to offer interactive exercises requiring independent reasoning and opportunities for manipulating data or physical variables that allow students to make discoveries for themselves.

- **The committee recommends aggressive investment in the development of innovative curricular materials for science education, with emphasis on data-driven interactive Web-based modules.**

A particularly fruitful area for educational software is the development of simple data-reduction tools and packaged exercises that will allow students to access and make insightful measurements from astronomical survey data. Interactive software running on an Internet browser employing standard tools such as Java could allow the user to make measurements of size, brightness, and color from astronomical survey images. From this experience, students can learn about classification, quantification, and experimentation and can experience the thrill and excitement of scientific discovery. There are many other possibilities in a field with the rich image-based heritage of astronomy. To ensure the pedagogical effectiveness of interactive software modules, it will be necessary to include an educator on each development team.

CONTRIBUTE TO A TECHNICALLY TRAINED WORK FORCE

Traditionally, many graduate departments in astronomy have focused their programs on careers centered on research, sometimes instilling in their students, consciously or not, a sense that alternate careers are less desirable and even carry a connotation of "failure." However, longitudinal studies conducted by the National Research Council's Office of Scientific and Engineering Personnel demonstrate that for several decades only about half of Ph.D. astronomers have ended up in positions where they identify themselves as being engaged primarily in research. However, nearly all recipients of the Ph.D. in astronomy are employed in science and engineering fields, with about 10 percent holding positions in industry (AAS, 1997).

It is difficult to anticipate whether these relatively stable long-term trends will continue, but there are two facts that must inform and guide astronomy graduate programs. One is that a Ph.D. in astronomy provides a versatile advanced degree in science, and the astronomical community needs to take more responsibility for ensuring that its students graduate prepared for a range of careers that require creative approaches to solving challenging technical problems. The second is that the total number of students pursuing and acquiring the Ph.D. in astronomy rose steadily during the past decade, reaching a total of 197 in 1997, up almost twofold from levels in the mid-1980s (NSF, 1999a).

In a year-long study discussing these issues and culminating in 1997 in *The American Astronomical Society's Examination of Graduate Education in Astronomy*, the community agreed that graduate education in astronomy should focus primarily on producing first-rate research scientists (AAS, 1997). Considerable support was also voiced, however, for broadening academic options that could lead to multiple career trajectories for students. It was also recognized that most graduate faculty are largely unaware of alternate career pathways that are available for students with training in astrophysics, and that these need to be identified.

Employment in industry is a career option that astronomers have exploited only rarely. There is a demand in industry for professionals who have broad training in basic astrophysics, but not the specialized knowledge associated with the research-based Ph.D. Some schools report that Ph.D. graduates encounter difficulty in marketing themselves to employers in industry, who seek a skill set that includes facility in project and database management, computational analysis, technical writing, and effective collaboration. In fact these skills are also required of successful professional astronomers, and departments need to identify ways to impart them more directly in their graduate curricula. The committee supports the recommendations of the 1997 AAS study.

- **The committee recommends that graduate programs in astronomy ensure that their students have the opportunity to acquire a broad range of technical skills that will enable them to pursue multiple career trajectories.**

The committee encourages some schools to develop enhanced or professional master's degree programs with applied astronomical connections, including instrumentation, computation, and education. For example, a professional master's program in applied astrophysics in cooperation with a local industry could provide a solid way to offer additional career choices. In addition, graduate departments should enable their students to make informed decisions about their career options by maintaining open records of statistics on employment and career paths for their Ph.D. and master's degree recipients.

Astronomy serves a national need, since it is the most effective of all the physical sciences in attracting students at all levels to a career in science and technology. The claim that astrophysics is a useful science does not end at the inlet of the pipeline. Surveys have shown that roughly half of those who earn Ph.D.s in astronomy and astrophysics take positions in other challenging fields. For example the computer industry, which is perennially short of qualified workers, values astronomers' and astrophysicists' expertise in and intensive experience with using computers. For similar reasons, graduates in astrophysics are attractive candidates for positions in financial engineering and econometrics, fields that emphasize highly mathematical and quantitative approaches to problem solving. Astronomers and astronomical image-processing techniques are an important resource for federal intelligence agencies concerned with national security. And the special-effects departments of the motion picture industry also borrow heavily from astronomy imagery and personnel. Directly and indirectly, astronomers and astrophysicists contribute to the national economy in far greater proportion than their numbers might at first glance suggest. A Graduate Assistance in Areas of National Need (GAANN) fellowship program in astronomy and astrophysics would be an excellent way for the nation to invest in a brighter economic future.

- **The committee recommends the inclusion of astronomy and astrophysics as a priority field for the GAANN fellowship program in the Department of Education.**

Issues regarding the training of a technically able work force may be even more relevant at the undergraduate level, where the interdisciplinary nature of astronomy could be more widely used to attract and prepare students for a wide range of scientific and technical careers. In fact, undergraduate programs in astronomy are not very common; fewer than 200 astronomy majors graduate nationally each year (AIP, 2000). The scarcity of undergraduate astronomy programs is probably due to the widespread feeling that future astronomers should major in physics as undergraduates and that concentration on astronomy should be postponed until graduate school. Although this approach serves the goal of training future researchers, it overlooks astronomy's potential as an undergraduate science major to offer students opportunities for strong connections to other disciplines, such as computer science or geology. In fact, astronomy is unique among the sciences in lacking a well-defined undergraduate curriculum. In contrast, many other scientific disciplines have recently held national discussions of undergraduate curricular reform, sharpening the goals and standards for undergraduate study in their fields.

- **The committee supports the efforts of the American Astronomical Society and the Astronomical Society of the Pacific to conduct national studies of undergraduate programs in astronomy to expand the broad-based technical education of science majors. The committee supports the efforts of these societies to coordinate effective interactions between the scientific and educational communities.**

The goal of these studies is to achieve community consensus on the desirability of expanding undergraduate programs in astronomy. Can the connection between astronomy and other scientific disciplines and the broad technical base inherent in astronomy and astrophysics provide a rationale for encouraging the growth of undergraduate programs in astronomy? Can such programs open up a wider range of scientific and technical careers for students?

PREPARE PROFESSIONAL ASTRONOMERS

Revolutionary advances in computing power, detector technology, and the technical sophistication of the new generation of telescopes are changing the face of astronomy. For most of the 20th century, astronomical research at the frontier was carried out by individuals or small teams of scientists. These teams worked with modest data sets or theoretical models. The advent of space observatories and large ground-based telescopes has changed this picture dramatically. Unprecedented requirements for technical support, the need for complex and expensive instrumentation, the growing requirements for comprehensive multi-wavelength surveys generating extraordinary quantities of digital data, the availability of unprecedented computing power, and the necessity for multinational collaborations all impose additional responsibilities on the programs that educate tomorrow's astronomers.

The continued success of the astronomical research enterprise requires that the astronomical community work to accomplish the following:

- Ensure the creation of the next generation of instrumentalists. The coming generation of telescopes will be outfitted with more sophisticated instrumentation than that seen in the past. The design and construction of these multimillion-dollar instruments will necessarily outpace the training of future instrumentalists in traditional Ph.D. programs. Special care must be taken to ensure both that graduate students have the opportunity to work creatively in all phases of instrumentation projects and that much of this work is done in research universities.
- Establish connections to other disciplines whose members have expertise in computational techniques, data-mining and algorithmic skills, and team and project management. Tomorrow's astronomers must be skilled in novel ways of manipulating and interpreting the large data sets that will inundate the field as the impending massive surveys are completed. Moreover, as the physical and financial scale of astronomical projects increases, the need for expertise in the design and management of large projects becomes more acute. Establishing links among graduate programs in other scientific or engineering disciplines facing these same issues will help to ensure that astronomy graduate students benefit from an interdisciplinary approach to acquiring these skills.
- Urge graduate departments (1) to instill a sense of responsibility in their students to contribute to public understanding of science in return for public investment in the research enterprise and (2) to expect their students to pursue careers in which research and education are more integrated than in the past.

EXISTING PROGRAMS AND FUTURE DIRECTIONS

Since publication of the 1991 survey, *The Decade of Discovery in Astronomy and Astrophysics* (NRC, 1991), scientists have become much more aware of the steps they can take to improve science education and of the opportunities available to pursue educational initiatives. Position papers such as *Science in the National Interest* (Executive Office of the President, 1994) sounded a "call to arms" to the scientific community to address science education. Responding to this call, the scientific community has recognized that integrating research and education is the best means of ensuring that our country both maintains world leadership in science and increases the public's scientific and technical literacy. In astronomy, the response has been dramatic, and scientists are eagerly participating in educational projects that aim to broaden science literacy, ranging from work with the nation's teachers to innovative curriculum development.

The NSF and NASA represent the two main sources of federal funding open to astronomers pursuing educational initiatives. The committee applauds the efforts made by both agencies to energize the scientific community in contributing to the reform of science education. The NSF's Directorate for Education and Human Resources and NASA's Office of Space Science have embarked on ambitious and exciting, if somewhat different, paths to enhance interactions between the science and the education communities. These programs are beginning to have a profound impact on science education and are a source of pride for both agencies. Maintaining momentum, however, will require that a number of concerns be addressed in this second decade of continued close connections between astronomical research and education.

At the NSF, most opportunities for educational initiatives are designed and administered through the Directorate for Education and Human Resources (EHR), while the majority of astronomers are more closely affiliated with the Astronomical Sciences (AST) Division in the Directorate for Mathematical and Physical Sciences. The opportunities within EHR were developed to address systemic means of improving science education, an essential aspect of NSF's mission. The majority of astronomers face two significant barriers to participation in NSF-sponsored educational projects. First, a previously defined structure with rigid guidelines leaves little room for submission of proposals conceived by astronomers. Second, the evaluation criteria of the research and education communities differ considerably. The NSF should encourage more cooperation between AST and EHR, with the objective of designing programs that address educational goals specific to astronomy. The NSF should ensure that astronomers as well as educators design and evaluate these programs.

NASA provides ample opportunities for astronomers to propose creative educational initiatives. One of the most popular programs, the NASA IDEAS grants, invites submission of a wide variety of educational initiatives from astronomers working in collaboration with educators, with the result that a broad spectrum of ideas flourishes. There is concern, however, that growing emphasis at OSS on strongly encouraging inclusion of an education and public outreach supplement in *all* research proposals may be misguided enthusiasm. A forced connection between research and outreach at the level of individual principal investigators can result in less than optimal results in both areas. Excellence in either research or education requires a significant commitment of time, and the principal investigator of a major research project may not be in a position to contribute to educational goals during the time period of the project. Moreover, without better agreement on what constitutes success in educational and public outreach endeavors, and better dissemination of positive results, there is a danger that even the most ardent astronomer may not make a lasting contribution through his or her educational initiative.

- **The committee urges greater communication between the federal agencies charged with increasing the participation of professional scientists in educational initiatives and with ensuring a healthy integration of research and education. A common set of goals, with a clearly articulated vision of how professional astronomers can make the most effective contribution to improving public science literacy, is needed. The goals, the clearly defined pathways for achieving them, and the standards for measuring success must be agreed on by both astronomers and educators.**

It is necessary in this decade to sharpen our understanding of what constitutes success in educational projects in astronomy, particularly the major ones. For large programs, such as NASA's OSS Educational Ecosystem initiative, it is essential that both the research and the education communities participate in evaluating projects and that, for those deemed successful, the outcomes be recognized by both communities as significant. Moreover, the elements of an educational initiative that have made it successful should be widely understood and emulated, and less successful efforts should also be publicized so that we can learn from past failures.

Finally, as a necessary step toward fostering a healthy climate in which both research and education efforts flourish and work synergistically, the astronomical community should encourage the leaders of academic institutions and federal laboratories to adopt a broadened reward structure for their staffs. This reward structure should explicitly acknowledge the importance of integrating research and teaching by recognizing the pursuit of high-quality educational initiatives along with excellence in research.

6

Policy for Astronomy and Astrophysics

INTRODUCTION

For well over 50 years the United States has enjoyed a leading position in astronomy. Remarkable studies of the skies with the Palomar 5-m telescope began in 1948. Rising to the challenge presented by Sputnik in 1957, the federal government put into place highly visible space- and ground-based programs. These marvelous resources for astronomy helped to attract some of the nation's best young minds to careers in science and engineering. The technological by-products of this effort, particularly in computing, aeronautics and astronautics, telecommunications, numerical simulation, and optics, have helped to give the nation an economic competitive advantage. The field of astronomy continues to attract scientists, and Ph.D. production is up. In 1987, 100 Ph.D.s were awarded, and in 1997 that number increased to 197 (NSF, 1999a). However, a critical time is at hand for astronomy in the United States. Space-based astronomy appears to be thriving, but U.S. leadership in astronomy as a whole is threatened by the decreasing share of federal investment in basic research in astronomy through the National Science Foundation (NSF).

The two lead agencies for astronomical research in the United States, NASA and NSF, support space- and ground-based studies, respectively. The Department of Energy (DOE) and the Department of Defense (DOD) also sponsor programs that include astrophysics. In the past decade, NASA and its scientists have been extraordinarily successful in communicating their scientific vision to the public and the Congress. Astronomy carried out in space, free from the interference of Earth's atmosphere, produces spectacular images of the cosmos at wavelengths ranging from the far- and near-infrared, through the optical and the ultraviolet, to the x-ray. Because "a picture is worth a thousand words," these beautiful and exotic images elicit a deep and immediate response among scientists and nonscientists alike. Their impact helps explain the public's enthusiasm for the nation's space program. The data provided by the suite of NASA missions has revolutionized our understanding of the universe.

Opportunities for U.S. astronomy from the ground using large optical and radio telescopes are equally challenging and compelling. For example, the Keck and Gemini telescopes offer high-resolution spectroscopic capabilities that, combined with theoretical analysis and computational modeling, can yield insight into the dynamics, chemical composition, and evolutionary state of the objects imaged from space as well as a wealth of other astronomical phenomena detected from the ground. In addition to very large filled apertures, another advantage that ground-based facilities have over their space-based counterparts is the short lead time between the latest breakthroughs in the fast-moving electronics and related industries, and the incorporation of such advances in sophisticated instrumentation at the back ends of telescopes. The generally much shorter time scale of ground-based projects also is better suited to the training of students. Ground-based facilities provide the stability of the long baselines required to produce images of faint sources at high angular resolution by interferometric techniques using arrays of telescopes. Large-scale optical, infrared, and radio surveys and synoptic studies, requiring decades of precise measurements on a large number of targets, may also be conducted advantageously from the ground. Adding to the excitement, ground-based astronomy is moving beyond traditional boundaries of optical and radio disciplines into neutrino and gamma-ray astronomy as well.

If U.S. astronomy is to remain world-class, improved resources for ground-based efforts must be provided. For ground-based astronomy, the NSF is the main source of federal money. Although the United States has been a world leader in astronomical research during much of the 20th century, other countries have advanced rapidly, so that in some cases their facilities are competitive with—and for some purposes, even superior to—U.S. facilities in optical ground-based astronomy.[1] The United States can benefit from international collaboration, but only if it brings world-class capabilities to the collaboration.

In this chapter, the committee recommends new policies to keep U.S. astronomy at the scientific frontiers. These recommendations were developed through the extensive efforts of the Panel on Astronomy Education and Policy. The chapter begins with recommendations related to the NSF, NASA, and the DOE, continues with comments on environmental factors affecting observing conditions for astronomy, briefly discusses professional development and the role of professional societies, and concludes with a reply to the questions posed to the committee by the Congress (see the preface).

POLICY RECOMMENDATIONS FOR THE NATIONAL SCIENCE FOUNDATION: GROUND-BASED FACILITIES

The NSF Division of Astronomical Sciences (AST) provides the National Optical Astronomy Observatories (NOAO), National Radio

Astronomy Observatory (NRAO), National Solar Observatory (NSO), National Astronomy and Ionosphere Center (NAIC), and Gemini Observatory facilities for the use of astronomers from the United States (and elsewhere). The committee commends the NSF for its role in giving U.S. astronomers the support needed to produce this current suite of world-class facilities, including the Very Large Array (VLA), the Very Long Baseline Array (VLBA), Gemini, and the Global Oscillations Network Group (GONG). In considering how to position the United States so that it can remain among the leaders in ground-based astronomical research in the future, it is helpful to recall recent history.

With respect to the funding of observing facilities, ground-based disciplines within astronomy have developed differently over the years. Major radio and solar facilities are now concentrated at national centers, whereas major optical facilities are concentrated at private and state (hereinafter "independent") observatories, which surpass the national optical facilities in aperture size and total telescope area. The radio community represents 10 percent of active astronomers; the much larger optical and infrared (OIR) ground-based community accounts for 40 percent of active astronomers. Both national and independent radio facilities provide observing time for the community at large. However, for ground-based OIR, about half the community has direct access only to the national facilities because their institutions cannot or do not participate in an independent observatory (NRC, 2000). To use a telescope at an independent observatory, these astronomers must forge a collaboration with someone who has direct access to the telescope.

In developing strategies for the new decade, the committee convened an ad hoc cross-panel working group chaired by F. Bash,[2] NSF-funded National Observatories, to review the functioning of all NSF-funded observatories and initiate discussion with the relevant panels and with the committee as a whole. Based on input from the Bash working group and on the work of both the education and policy panel and the Panel on Optical and Infrared Astronomy from the Ground, the committee recommends a new paradigm for ground-based astronomy that it believes will lead to the most effective use of ground-based facilities and optimize the science opportunities for the astronomical community. The committee then outlines the roles and responsibilities of the national astronomy organizations and the independent observatories, as well as those of the NSF, in this new paradigm.

RECOMMENDED NEW PARADIGM

The United States has a long tradition of independent optical observatories, beginning more than 150 years ago. Construction of the first U.S. observatory, located at Williams College, was completed by 1838; citizens of Boston donated funds to acquire a twin of the world's largest refractor—a 15-inch telescope—for Harvard College in 1847. Most recently, the Keck Foundation provided the bulk of the funding for two 10-m telescopes for a consortium led by the University of California and Caltech. Currently, the national facilities have 22 percent of the total area of primary mirrors of U.S. optical telescopes, and the independent observatories have the dominant 78 percent (ESO, 1998). The suite of national observatories that serves the U.S. astronomical community—NOAO, NRAO, NAIC, and NSO—was created during the period from 1957 to 1983; Gemini was formed in 1993. Each was created as a result of the arguments put forward by particular scientific communities at different times and in different contexts. Not surprisingly, each has evolved in its own way, and each has enjoyed different successes.

As we enter the new millennium, U.S. ground-based astronomy facilities, both independent and national, must evolve in a changing environment. There is growing competition from Europe and Japan, which together have invested in OIR facilities at a level (relative to gross domestic product) greater than 10 times that of the NSF investment over a comparable period,[3] and more than 3 times that of the combined federal, state, and private investment. The large investment of state and private funding in major OIR facilities provides the context and opportunity for using federal funds in a highly leveraged way to ensure that the distributed system of U.S. ground-based facilities as a whole has the capabilities to compete successfully for a world leadership position in all of astronomy. Universities, which will produce the next generation of astronomers and instrument builders, as well as other scientists and engineers, must be supported as a key part of the new system. Finally, the public that provides a substantial part of the support for the system must be informed about exciting astronomical discoveries. The committee believes that by working in concert, the independent observatories and the national facilities can ensure that astronomy in the United States will thrive and move forward to capture the scientific opportunities ahead.

- **To help ensure maximum scientific return from federal investments in ground-based astronomy, the committee recommends that all facilities, whether nationally or**

independently operated, be viewed as single integrated systems—one for optical and infrared astronomy, one for radio astronomy, and one for solar astronomy. The committee recommends that the NSF Division of Astronomical Sciences implement a plan for ground-based astronomy that reflects an integrated view of independent and national observatories and the funding available from government and private sources.

ROLES AND RESPONSIBILITIES OF NATIONAL ASTRONOMY ORGANIZATIONS AND INDEPENDENT OBSERVATORIES

To move forward to the next generation of facilities, which are likely to be of a scale that will require a collaborative approach, the committee envisions the following as the responsibilities of the participants:

1. Community participation in major national telescope initiatives must be led by an effective national astronomy organization working in concert with universities and similar institutions. Such an organization should in turn be subject to close community oversight with appropriate advisory bodies. It should:

- Lead the development of a strategic plan for the evolution of the capabilities of the system by organizing discussions involving the NSF, the independent observatories, the academic community, and industry.
- Be able to contribute to the scientific leadership and provide the technical expertise (e.g., professional engineering and system management), the administrative skills, and the management experience and infrastructure needed in the building of those facilities that are too large or expensive to fit within the resources of single institutions or small partnerships.
- Ensure that the United States enters international collaborations with a clear scientific purpose and a well-considered technical and administrative approach, and maintain these or modify them as appropriate for the duration of the project.
- Coordinate with the community to provide capabilities that support the suite of state-of-the-art large telescopes; such capabilities may include telescopes, instruments, archives, observing modes, and other channels for access to data.

- Collaborate with universities to build instruments for national telescopes with agreed-upon and clearly documented technical standards.
- Establish internships for instrument builders at national observatories in order to foster the training of skilled instrumentalists to benefit both U.S. astronomy and U.S. industry.
- In the case of the national OIR organization, administer publicly available observing time at federally funded telescopes such as Gemini and, where appropriate, the publicly available time at the independent optical observatories.

2. Recognizing that scientific progress will be strengthened by a cooperative approach on the part of the national and independent facilities, universities/independent observatories should:

- Develop acceptable mechanisms in concert with the NSF and the relevant national astronomy organization for sharing fractions of their facilities with the larger astronomical community. The Telescope System Instrumentation Program, presented in more detail in Chapter 3 of this report and in Chapter 2 of *Astronomy and Astrophysics in the New Millennium: Panel Reports* (NRC, 2001), is a prime example of such a mechanism. For the independent observatories, this scheme has the advantage that (a) no one facility will have to provide every capability since diverse facilities will be open to all and (b) their scientific staffs will benefit through increased access to other facilities.
- Work with the appropriate national astronomy organization to develop a strategic plan for the system as a whole, and implement the parts of the plan that should be carried out at the universities and independent observatories.
- Work with the national astronomy organization on the development of facilities that are too large or expensive to fit within the resources of single institutions or consortia.
- Assume the responsibility for purchasing, instrumenting, and operating *small* telescopes needed for their students and faculty.

3. The committee's review, reflected in its policy recommendations, led to the following assessment of the current national astronomy organizations:

- NOAO as currently functioning and overseen is not structured to fulfill the role foreseen for an effective national organization acting on behalf of the ground-based OIR community. The committee believes that NOAO's goals and operations must be substantially modified to conform to this paradigm. The NSF, together with NOAO and the Association of Universities for Research in Astronomy (AURA), should establish a common vision of how these new roles can be implemented, and the NSF should set criteria, based on the precepts listed under 1 above, by which NOAO's success can be evaluated. Using these criteria, the NSF should initiate a high-level external review of NOAO to ensure that changes to achieve these goals can be instituted promptly. Periodic reviews will keep this transition and further evolution on course.

- Gemini promises to provide the U.S. astronomical community with two telescopes consistent with the highest recommendation of the 1991 survey, *The Decade of Discovery in Astronomy and Astrophysics* (NRC, 1991). NOAO deserves considerable credit for conceiving and promoting the Gemini project and for supplying much of the technical capability that has made it possible. However, NOAO must continue to exert leadership through the U.S. Gemini Program and, in concert with the U.S. members on the Gemini Board, marshal the expertise to design and build future Gemini instruments that will respond to the science goals of the U.S. community.
- NRAO has won the respect of the radio community. It should continue to engage the broad university community in developing facilities and instruments, and it should work proactively to ensure that radio astronomy science and instrumentation development are firmly rooted in the universities.
- NAIC is an example of a university based and operated national observatory that is a uniquely powerful facility. Arecibo serves as an excellent model by inviting the academic community to operate university-built receivers on the telescope.
- NSO has developed world-leading research capabilities that solidly support both the U.S. and international community of solar astronomers. The committee believes that NSO and NOAO will be well served by NSO's separation from NOAO in order to accomplish the next solar telescope initiative. The committee is pleased to see that this separation has begun.

NEW PROCEDURES AND STRATEGIES

If U.S. astronomy is to step up to new challenges, embrace expanding scientific opportunities, including international collaborations, increase the power of its facilities (with concomitant increases in size and complexity), and exploit the many advances in technology, it is essential to make optimum use of government funding and capitalize on private investment. The committee recommends five strategies for the National Science Foundation.

1. Competitive Review of NSF Astronomy Facilities and Organizations

- **The committee recommends that, about every 5 years, the National Science Foundation astronomy facilities be competitively reviewed and prioritized based on past performance and future expectations. A single committee should conduct this review across all subfields of astronomy. New facilities should first be competitively evaluated between 5 and 10 years after they become operational and on the normal cycle thereafter.**

Ground-based astronomy telescopes generally have useful lifetimes of at least 15 years or more because they can be modified with state-of-the-art equipment to approach their highest possible levels of performance. Yet, as in all rapidly developing fields, with changing scientific questions and opportunities, some telescopes remain more appropriate than others for addressing new scientific challenges. In evaluating how best to advance, the astronomy and astrophysics community must therefore determine not only which new facilities should be built, but also those that should be modified, privatized, or even shut down.

The committee suggests that all NSF astronomy facilities be subjected to a competitive review, with prioritization and privatization or closure as possible options. NASA carries on such a review, called the Senior Review, on a 2-year timetable, which will move to a 3-year cycle in 2003. This review cycle is appropriate to the shorter life of space missions. After the Senior Review, highly ranked missions benefit from increased operations and grants funds; the lowest-ranked missions are frequently ramped down and turned off. Such a procedure is designed to maintain the best science and also to make funds available for new or improved facilities or grants. The committee conducting the competitive review for the NSF may wish to request comparative evaluations of facilities within individual disciplines prior to its deliberations. The competitive reviews should be timed so as to provide assessments helpful to the NRC's decadal surveys of astronomy and astrophysics.

2. Regular Expert Advice for the NSF Division of Astronomical Sciences

The competitive review procedure will offer advice to NSF AST on facilities and programs at 5-year intervals. The NRC's Committee on Astronomy and Astrophysics provides strategic advice and oversight for the implementation of the decadal survey. NSF AST gathers community input in a variety of formal and informal ways. Beyond these sources of advice, NSF AST has a critical need for regular expert advice on pressing issues that arise, both in the short and long term. The committee strongly encourages NSF AST to find a mechanism to obtain such advice, on a continuing basis, from a small group of experts representative of the community.

3. Funding Attached to New Facilities

- **The committee recommends that the National Science Foundation budget for any new capital project include funding for operations, new instrumentation development, and research grants associated with the new facility, as well as for construction and the initial complement of instruments. This combined allocation must be regarded as the cost of the new facility. The committee emphasizes that the inclusion of grant money specific to each new facility should not displace the existing NSF Division of Astronomical Sciences unrestricted grants program.**

Compelling scientific challenges drive the construction of new facilities for ground-based astronomy. Hardware alone does not ensure a successful facility; reaping the scientific benefits a facility can provide requires substantial complementary activities. These efforts include broadly based observing programs, frequently at different wavelength bands, challenging theoretical research, and ongoing development of instrumentation to benefit from advances in new technologies, instrument design, and computational power. Starting construction without an overall budget in hand for a complete program can spell lost opportunities for researchers who could capitalize on the powerful new capabilities; bare-bones instrumentation efforts unable to move forward with technology developments; and operations funds inadequate to realize the scientific potential of the facility.

The committee strongly recommends that a different mode of budgeting be instituted. The community and the NSF must commit *at the beginning of a project* to estimate adequately and acquire the funds required to properly utilize the facility for the research of which it is capable. Without clear identification of these monies, new construction should not begin. This committee rates the importance of the unrestricted grants program so highly that it emphasizes that *new initiatives should not be undertaken at the expense of the unrestricted grants program.* Based on the Panel on Astronomy Education and Policy's examination of the costs of existing radio and optical facilities, the committee estimates that each year about 5 to 10 percent of the total cost of capital construction—materials and labor—is required to support reliable operation for a full range of observing modes. About 3 percent per year will be needed for the first 5 years of operation to upgrade the initial instrumentation and to provide a suite of new instruments that fully exploit the capabilities of the new facility. These estimates are necessarily approximate because different facilities count faculty and staff positions in different ways, and the cost of a facility may or may not include the cost of site development or base camps. The 3 percent figure for instrumentation is somewhat higher than the average for existing facilities because instrumentation is generally underfunded. For example, had this recommendation been in place for the past decade, the scientific capability of the VLA would not be compromised by the use of technology left over from the 1970s.

To fully exploit new facilities, researchers should be able to carry out ambitious, creative programs that develop the facilities' full capabilities. The committee therefore recommends that an additional 3 percent of the capital cost be budgeted for each of the first 5 years for facility grants for major ground-based facilities. For moderate ground-based facilities, a cost-effective and competitive grants program requires a somewhat higher percentage, and the committee recommends 5 percent per year for such facilities. The funding for small ground-based facilities is inadequate to justify facility grants programs. Facility grants for the major and moderate facilities would support the costs of the users of the facility, research grants related to the use of the facility, including multiwavelength studies, and support for the theory challenges associated with the facility. As is the case for individual investigator grants, facility grants should be allocated in open competition in the national community.

In estimating the total cost of ground-based capital projects, the committee has adopted the figures of 7 percent per year for operations, 3 percent per year for new instrumentation, 3 percent per year for grants for major facilities, and 5 percent per year for grants for moderate facilities. The cost estimates include this additional funding for a period of 5 years (with the few exceptions detailed in Chapter 1), when the scientific returns per dollar invested are at their peak. Following this 5-year period, the mix of operations, instrumentation, and grants funding should be subject to the recommended competitive review to ensure that additional investments in the facility are balanced against requests for new facilities and for new instrumentation and grants for other facilities. As a guideline, the committee anticipates that the funding for facility grants will be reduced substantially after 5 years. Operations and instrumentation funds are likely to ramp down more slowly.

4. NSF Responsibilities for Access and Balance in the System of Independent and National Observatories

- **To help ensure optimum scientific returns from the system of independent and national observatories, the committee recommends that the National Science Foundation enhance and leverage observing opportunities for the community and that it recognize the interdependent roles of universities and national facilities.**

The proposed new paradigm for ground-based astronomy envisions the suite of U.S. astronomical facilities, including federally supported and independent observatories, as single systems for each subdiscipline of ground-based astronomy: OIR, radio, and solar. Within a system, each facility will inevitably have its own particular capabilities, and U.S. observers should, in principle, be able to apply for observing time with the instrument best suited to the problem at hand. The committee encourages the NSF AST to adopt a flexible approach to ensure community access to these observational resources coupled with adequate development of young scientists. The approach may vary from one branch of ground-based astronomy to another.

For OIR astronomy, a major issue involves community access to the telescopes of independent observatories. In the United States, a number of the large optical telescopes constructed recently have been built with private and/or state funds. Access to most of these independent facilities is relatively restricted. The committee's recommendation in Chapter 3 for the Telescope System Instrumentation Program can foster open, competitive access to these powerful new facilities. In the proposed program, construction of new instrumentation at private observatories is funded in exchange for telescope time or other, equally valuable benefits for the community at large.

The situation for radio astronomy is currently close to the concept of one system. For millimeter-wave astronomers, in particular, there is significant access[4] to the facilities operated by universities. However, as the Panel on Radio and Submillimeter-Wave Astronomy points out in Chapter 4 of the *Panel Reports* (NRC, 2001), the dominance of the national centimeter-wave facilities has reduced the opportunities for contributions by university radio astronomy in this area. With GBT and ALMA coming on line in this decade, it will be critical to ensure that radio astronomy continues to be carried out in a university setting. The recommended initiatives CARMA and SKA technology development can and should continue to involve university astronomers.

For solar physics, for which most of the major facilities are national facilities, it is important to ensure vigorous university programs so that young solar physicists can be trained. Solar physics is currently very healthy given the success of recent space-based missions. However, much of solar physics is now carried out in research institutes, and university departments are not replacing solar physics faculty as they retire. The NSF should work to involve university astronomers both in the planning of new solar facilities and in solar physics research programs. In particular, the committee recommends that the NSF encourage strong university participation in both the Advanced Solar Telescope and the Frequency Agile Solar Radio telescope initiatives to provide opportunities for developing the next generation of solar physicists.

5. Management Oversight

- **For National Science Foundation-sponsored projects costing several million dollars or more, the committee recommends that NSF require a management plan appropriate to**

the size of the project, with suitable oversight mechanisms to ensure successful completion of such projects on time and within budget.

Over the past decade, ground-based astronomy in the United States entered a new era in terms of the size and complexity of its new facilities, instruments, and software. The costs of major facilities now often exceed $100 million, while the costs of facility instruments can approach $10 million. The talents of many people are needed to successfully complete such programs, requiring the application of management approaches and tools appropriate to projects of such scale. The evaluation of the management plans for NSF proposals for facilities and instruments costing more than several million dollars should constitute a significant part of the proposal selection process. Continued oversight should take place through post-award reviews of the progress made toward planned objectives. Project proposals should contain contingency plans for a reduction in scope should costs begin to escalate beyond the original estimates. The committee recognizes and approves of the steps that the NSF has already taken in this direction.

NATIONAL SCIENCE FOUNDATION GRANTS IN ASTRONOMY AND ASTROPHYSICS

A severe problem currently exists in the budget levels for ground-based astronomy. Most of the NSF's support for astronomers comes through the Division of Astronomical Sciences, while some funding comes through the Division of Physics. NSF support for astronomy and astrophysics has not shared proportionately in the budget increases of the NSF as a whole (Figure 6.1). Consequently, the strains of decreased purchasing power coupled with the responsibility for maintaining the national U.S. ground-based effort in radio, optical, and solar astronomy have led to a division of the NSF AST budget such that, through the decade of the 1990s, about 65 percent of the NSF allocation to AST was assigned to facilities operated by national astronomy organizations, with only about 22 percent made available to support individual investigators (Figure 6.2); the rest went to instrumentation and the university radio observatories. The budgets for the national facilities have lost purchasing power, resulting in cutbacks of services to users and reductions in necessary maintenance, and sharply reducing resources for improve-

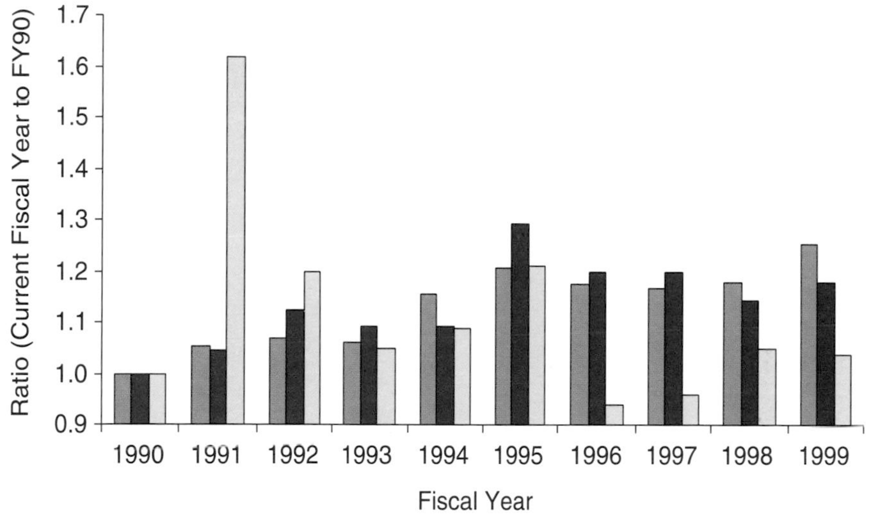

FIGURE 6.1 Budget history, as ratio of as-spent dollars for FY1991 through FY1999 to spending for FY1990, for National Science Foundation R&D (blue); NSF's Mathematics and Physical Sciences Directorate, including spending for major research equipment (red); and NSF's Division of Astronomical Sciences, including funding for astrophysics from NSF's Division of Physics, and spending for major research equipment (yellow).

ments in these national facilities. The ratio of proposals received by NSF AST to proposals funded in the university grants program doubled from 2:1 in FY1990 to nearly 4:1 in FY1999 (see Figure 6.2). The corresponding ratio for the Mathematics and Physical Sciences (MPS) Directorate and the NSF as a whole was more favorable at 3:1. More astronomers are applying to the NSF for support; between 1990 and 1999, the number of proposals received increased by about 50 percent, and the number of grants awarded declined by 30 percent (NRC, 2000, Figure 5.4).

Individual investigator grants support the work of astronomers who propose in peer-reviewed competition to develop innovative theoretical ideas and challenging observational programs. The studies funded by these grants form the fabric of basic research in astronomy and provide a primary means of training graduate students, and they provide some support for undergraduates as well. During the past decade, the shrinking opportunities for research grants in astronomy have had severe effects on the university research community and appear out of balance with respect to other divisions within the MPS. Restrictions in NSF

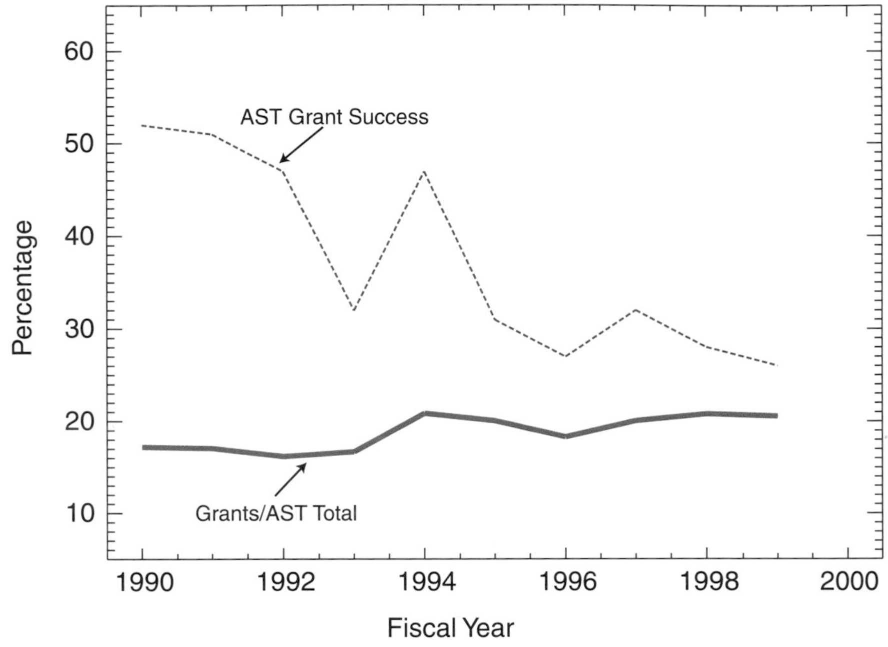

FIGURE 6.2 Percentage of National Science Foundation Division of Astronomical Sciences competitively reviewed individual investigator grant proposals funded (top curve) and funds in these grants as a percentage of the total Division of Astronomical Sciences budget (bottom curve), FY1990 to FY1999.

funding have compromised the ability of U.S. researchers to obtain the full scientific return on the funds invested in ground-based facilities. Although many astronomers have responded by seeking funding from NASA, the science that they can do is altered as a result and their funding is vulnerable to the failure of one of NASA's flagship missions (NRC, 2000, Finding 4, p. 2). To improve the picture, the committee recommends two approaches to enhancing the NSF AST grants program:

- With construction of each new facility, commit to a facility grants program, including it in a comprehensive project budget along with the allocations for construction, operations, and instrumentation.
- Through a periodic competitive review of NSF-funded facilities, make existing funds available for grants or for new or improved facilities, depending on scientific priorities.

In addition, the committee recommends the following :

- To ensure the future vitality of the field, NSF AST should provide adequate funding for grants not tied to a facility or program—the unrestricted grants.
- The NSF AST should initiate new projects with other NSF divisions that emphasize cross-disciplinary activities deriving from new technologics or new scientific themes developed in this report.
- The NSF should work with other federal government agencies and with the astronomical community to build interagency programs that will aggressively pursue astronomical problems of broad national interest.

The committee notes that the growth of interdisciplinary projects in astrophysics and cosmology has led to increased participation of NSF divisions other than AST in support of astronomy and astrophysics. In particular, the committee endorses the continued development of the new program activity coordinated within the NSF Physics Division (PHY) for projects in nuclear and particle astrophysics. This initiative cuts across NSF programs in high-energy physics, nuclear physics, theory, and gravitational physics within PHY as well as work in the Office of Polar Programs and NSF AST. It represents a positive step in developing a coherent approach to interdisciplinary projects in astrophysics within NSF.

POLICY RECOMMENDATIONS FOR NASA: SPACE-BASED ASTRONOMY

The U.S. space program in astronomy is generally vigorous and strong. NASA has made a variety of flight and suborbital opportunities available to the astronomy community along with grants for data analysis and theoretical studies. NASA's initiative for the Astrophysics Data System has vastly increased the accessibility of the scientific literature for astronomers. NASA deserves credit for this valuable activity and is urged to continue it. NASA's data archiving and distribution systems are extremely effective as well, making results from space-based observations available to a worldwide community of scientists. NASA's efforts to engage the American public in the excitement of astronomical search and discovery are exemplary.

As an agency, NASA has been successful in communicating its

objectives to the Congress and the community at large in terms of "themes" as described by its Office of Space Science. The Origins theme has been particularly resonant with Congress. The agency is to be commended for its proactive approach in clarifying its goals and in changing its research and analysis policies to attempt to ensure optimum science returns from the investment in missions.

The committee is concerned, however, that NASA maintain the diversity in mission size, from small to medium to large, needed to meet scientific objectives in a cost-effective manner.

- **The committee recommends that NASA maintain diversity in its flight programs by ensuring that a suite of opportunities, including small, moderate, and major missions, is available to accomplish scientific goals.**

There are compelling scientific, programmatic, technical, and educational reasons for ensuring some balance between the major flagship missions and missions of moderate and small size (and cost). Flagship missions such as the Compton Gamma Ray Observatory, Hubble Space Telescope, and Chandra X-ray Observatory occur about twice per decade and produce outstanding science that defines substantive new areas of research in astronomy. But because of their high public visibility and great costs, such missions must be designed and built so as to maximize the likelihood of operational success in orbit. Ensuring a high success rate tends to drive up costs and lengthen schedules. The future of astronomy in space will be at substantial risk if it must depend on the successful deployment of only a few missions per decade.

Small and moderate missions add important dimensions to NASA's space astronomy program: respectively, rapid response and targeted science. The Explorer program, an effective response to the need for frequent small-mission opportunities, should be continued at its current level. Because they can deploy new technology on relatively short time scales or move rapidly to follow up on recent discoveries, newly conceived missions of moderate cost can at times scientifically outperform the large missions on particular problems. Given the lower costs of small and moderate missions, an occasional failure can be accepted, although no failure in space occurs without some political cost to the program. Compared with one or two larger missions, several moderate and many small missions will more likely provide greater opportunities for developing a diverse set of new technologies and for training experimental space scientists. On several accounts, moderate and small missions can be extremely cost-effective.

The committee believes that a vigorous program of moderate-sized missions is required to achieve program diversity. Moderate missions are similar in capability to the older Delta-class Explorers such as the Cosmic Background Explorer, the Infrared Astronomical Satellite, the International Ultraviolet Explorer, and the Rossi X-ray Timing Explorer or to the Discovery missions of the Solar System Exploration and Discovery Program. Several missions recommended for this decade by the committee fit into this class: ARISE, EXIST, GLAST, LISA, and SDO.

The committee also calls attention to the potential for very cost-effective and scientifically fruitful advances with the advent of long-duration (10-day) balloon flights and the expected availability after the final 2 years of development of ultralong-duration (up to 100-day) balloon flights (see the section "The National Virtual Observatory and Other High-Leverage, Small Initiatives" in Chapter 3). These balloon-borne scientific missions also promise nearly ideal opportunities to involve students in all aspects of a flight program from creation of the idea through analysis of the data.

POLICY RECOMMENDATION FOR THE DEPARTMENT OF ENERGY: ASTROPHYSICAL RESEARCH

- **Given the increasing involvement of the Department of Energy in projects that involve astrophysics, the committee recommends that DOE develop a strategic plan for astrophysics that would lend programmatic coherence and facilitate coordination and cooperation with other agencies on science of mutual interest.**

Particle and nuclear astrophysics and cosmology are branches of physics. Because of the extreme scales of distance and energy accessible in the universe, astronomical observations can probe particles and forces in ways not possible in the laboratory. As a consequence, DOE-supported high-energy and nuclear physicists and terrestrial laboratories are increasingly making contributions to astronomy and astrophysics through instrumentation, detectors, techniques for acquisition and analysis of data, and numerical simulation. A highly visible manifestation is the Los Alamos preprint server (see <xxx.lanl.gov>), which has become a valuable and widely used resource for dissemination of research in astronomy as well as in high-energy, nuclear, and gravitational physics.

As the size and complexity of astronomy and astrophysics projects increase, funding patterns are changing in ways that challenge traditional agency boundaries and funding patterns, and interagency collaborations are frequently advantageous. The committee commends DOE for supporting astrophysical research and recommends that DOE develop a strategic plan for astrophysics to ensure a vigorous, coherent research program and to facilitate cooperation with other agencies.

ENVIRONMENTAL IMPACT ON ASTRONOMICAL OBSERVATIONS

Observing conditions for astronomy, both on Earth and in space, are deteriorating—in many instances quite rapidly. Many new initiatives presented in this report may never reach their full potential unless skies are dark, space trash is kept under control, and unwanted radio emissions are kept in check. Viewing the beauty of the night skies is among the most awe-inspiring of human experiences, and all people should be concerned that pollution is destroying that view.

Preventing an adverse environmental impact on dark skies must generally be addressed on a local level with judicious siting and communication with neighbors, local establishments, and local governments. Building awareness of the problems of light pollution and of possible solutions for achieving quality outdoor nighttime lighting will improve the nighttime environment for everyone—and generally reduce energy costs, too.

Global light sources proposed for space are an alarming prospect. Protecting against such worldwide incursions will require the cooperation of national and international professional societies as well as governments. The International Dark Sky Association is to be commended for bringing important issues like these before the public and supporting efforts to teach the principles of good lighting on the ground.

Posing yet another threat to astronomical observations, space debris generated by satellite launches, human activity in space, and inoperative hardware can become a major hazard for telescopes in space. NASA and the professional societies should work toward developing international protocols to address this issue.

In addition, pressure toward increased commercial use of the electromagnetic spectrum is growing. The past two decades have seen a huge increase in the number of end users of already-popular applications, such as cell phones and the Global Positioning System, and an enormous variety of new applications continue to be introduced. The result has been significant contamination of much of the frequency space with unpredictable and broadband emissions from an array of communication devices. Although many applications of the radio spectrum provide a clear benefit for society, concern is growing about protecting observing conditions for radio astronomy, a uniquely powerful tool for studying the universe. The committee supports the efforts to deal with these important problems in several ways: better community relations and public information on the pollution problems; negotiations with violators on a case-by-case basis; and increased research oriented toward making radio telescopes less susceptible to interference (e.g., adaptive cancellation, filter technology, and high-spectral-efficiency modulation techniques). Specific technical issues are discussed in Chapters 4 (on radio astronomy) and 2 (on OIR astronomy) in the *Panel Reports* (NRC, 2001).

ISSUES OF PROFESSIONAL DEVELOPMENT

The past decade saw an increase in the contingent of professional astronomers in the United States. The Space Telescope Science Institute's (STScI's) 1999 survey of astronomy (discussed in AAS CSWA, 2000) documented substantial growth in the field between 1992 and 1999—a 33 percent increase in the number of Ph.D. astronomers active in 32 U.S. departments of astronomy and in four observatories with equivalent science facilities. The number of astronomy Ph.D. recipients in the United States increased by 37 percent over the period from 1994 to 1997 to reach a total of 197 graduates in 1997 (NRC, 2000, Table G.1). This growth, reflecting the high interest of students in the astronomical sciences and the job opportunities that exist, contrasts with the declining number of graduates in physics (down by 11 percent) and chemistry (down by 6 percent) over the same period (NSF, 1999b).

A growing field presents opportunities to address areas in which professional development can be enhanced. On these subjects, the committee received many helpful communications from the community. Praise was high for the NASA/STScI Hubble Fellowship program and the more recent, highly selective fellowships associated with the large NASA missions CGRO and Chandra. The recently instituted NSF CAREER program for young faculty appears to attract strong candidates who wish to integrate educational activities with their research and other professional responsibilities.

Several issues in professional training and development are worthy of comment: postdoctoral training, NASA's Long-Term Space Astrophysics program, and the participation of women and minorities.

POSTDOCTORAL TRAINING

Postdoctoral fellowships play a critical role in the career path of most researchers in astronomy and astrophysics. Despite the great success and high visibility of the Hubble Fellowship program over the past decade, enhanced support of postdoctoral associates—both grant-funded and portable—is needed in many areas of astronomy and astrophysics, including ground-based astronomy, instrumentation, and theory. Given the investment in new facilities, it is vital to encourage talented young people to become involved in instrumentation. Since substantial resources are needed to build new instruments, the committee recommends that postdoctoral positions for instrumentation be associated with particular projects or facilities to ensure that the needed resources will be available.

The committee recommends a targeted program of portable postdoctoral fellowships in theoretical astrophysics, jointly supported by NASA and the NSF, as a high-leverage small initiative. The focus on theory reflects the committee's strong belief that its ambitious recommended program of new facilities will be most effective if a small fraction of the investment is used to support talented young researchers who will expand—or perhaps tear down—the theoretical framework on which the design of these missions is predicated.

NASA'S LONG-TERM SPACE ASTROPHYSICS PROGRAM

One opportunity exists at NASA to receive research support over a 5-year period: the Long-Term Space Astrophysics (LTSA) program. Continuity of support for this period is highly regarded by the science community because it enables research of adequate depth on a substantive issue in astronomy and astrophysics. A certain fraction of the grants has traditionally been set aside for scientists in the first stages of their careers so that they might establish a record of independent research in preparation for permanent or tenure-track positions. One study of the progress of junior LTSA scientists, however, indicated that this goal has not, in fact, been achieved:[5] Junior LTSA scientists have been substantially less successful in obtaining tenure-track faculty positions than their peers with named fellowships, such as the Hubble fellowships. The committee therefore recommends that in the LTSA program the separation between senior and junior scientists be eliminated, and that the selection of grants be based solely on scientific merit.

WOMEN IN ASTRONOMY

In 1999 women constituted 21 percent of the active membership of the American Astronomical Society (AAS). Each year between 1920 and 1995, women who earned Ph.D. degrees in astronomy represented 8 to 20 percent of the total astronomy Ph.D.s awarded (AAS CSWA, 1996; AAS, 1997). In the competition for NSF grants in astronomy over the decade from 1988 to 1997, the success rate of women exceeded that of men for 8 of the years. Assessment of the representation of women at various professional levels is based on the assumption that the pool of women astronomers corresponds to the percentages documented over the years for astronomy Ph.D. production. The 1999 STScI survey suggests that in 1999 a lower percentage of women (41 percent) than men (58 percent) advanced from graduate school to a postdoctoral position. Comparison by gender of those holding postdoctoral appointments in 1992 with those holding junior faculty appointments in 1999 suggests that men and women were selected at similar rates for entry-level faculty positions. But women's advancement to higher positions is no faster and apparently slower than men's. Both the 1999 STScI survey and a 1999 AAS demographic survey (discussed in AAS CSWA, 2000) show that for astronomy faculty in U.S. institutions, women constitute 18 percent of the assistant professors, 13 percent of the associate professors, and only 6 percent of the full professors. While the fraction of women at the assistant and associate professor level reflects the Ph.D. production rate, that at the full professor level does not.

Astronomy is a relatively small discipline, making statistical study of the field's demographics difficult. There are only 30 women astronomy professors at 34 universities. However, studies across the physical sciences that include astronomy consistently show that men advance to senior positions more rapidly than women, even when starting from the same point, such as a prestigious NSF or NRC postdoctoral fellowship. This disparity does not appear to stem from any of the sociological factors that might distinguish women from men in current society; rather, the prevailing model is that women suffer from an accumulation of smaller disadvantages, which together result in longer time to tenure or to promotion to a full professorship, less pay compared with that for men who have similar credentials, and diminished representation at the top echelons of scientific society (Sonnert and Holton, 1996; Valian, 1998). A report on senior women on the Massachusetts Institute of Technology (MIT) science faculty concluded that gender bias prevailed on that campus. The dean of science at MIT acknowledged that gender discrimination against senior women science faculty marginalized the women, undervalued their achievements, and excluded them from positions of power (MIT, 1999).

The astronomy community needs to understand why women are underrepresented in certain areas, and thorough studies of career patterns should be continued. The committee calls on the community, particularly those in leadership positions, to ensure equitable treatment of women in astronomy.

MINORITY SCIENTISTS IN ASTRONOMY

Blacks, Hispanics, and Native Americans are underrepresented in the total U.S. science and engineering labor force. Blacks and Hispanics constitute 19 percent of the population and the total labor force, but only 8 percent of the science and engineering labor force (NSF, 1994).[6] Furthermore, these groups constitute only 6 percent of science and engineering Ph.D. recipients in the United States, and less than 5 percent of those employed in academia (NSB, 1996; NSF, 1996, Table 3). Although the participation of Blacks, Hispanics, and Native Americans in advanced high school mathematics classes increased between 1982 and 1994, their scores in standardized mathematics tests were still lower than those of other students, and the discrepancy did not diminish between 1990 and 1996 (NCES, 1996). Mathematics and science achievement at the K-12 level is critically important as a basis for further science and engineering study.

Currently, these minorities account for 4 percent of the AAS membership.[7] Between 1986 and 1995, the fraction of astronomy Ph.D.s awarded to minority scientists ranged between 1 and 4 percent; over a 22-year period the average was 2 percent.[8] The success rate of minority principal investigators in winning NSF AST grants is comparable to the rates for all women and men over the past 10-year period—minority scientists appear fully competitive in winning NSF grants.

To identify areas of concern and to foster mentoring and other opportunities for underrepresented minorities at all levels of experience, the AAS has established the Committee on the Status of Minorities. The importance of a strong K-12 mathematics and science education in fostering scientific and technical careers represents an opportunity for well-designed programs in astronomy to have a positive impact in attracting minorities to science. Astronomers could be significant role models and mentors for young minority scientists and students. The committee believes that providing all members of the community equal access to professional opportunities will yield the strongest science. Achieving this goal will require the efforts and the support of all members of the astronomical community.

ROLE OF PROFESSIONAL SOCIETIES

The major astronomical societies active in the United States are the American Astronomical Society, the Astronomical Society of the Pacific, and the Division of Astrophysics of the American Physical Society. They differ in their membership, activities, and goals, but each offers a valued resource to the astronomical community.

The American Astronomical Society, the society for professional astronomers in North America, contributes substantially to the progress of astronomy by publishing scholarly journals, organizing scientific meetings, and, more recently, expanding its focus on astronomers' education and employment needs. Among all the scientific societies, the AAS has clearly taken the lead in communicating the results of its meetings to the public at large and has paved the way for the specialized divisions to follow suit. Responsive to the needs of the community, the AAS reviews and issues small research and travel grants (sponsored by the AAS, NASA, and the NSF), holds town meetings with agency representatives, and has recently demonstrated its leadership in preparing *The American Astronomical Society's Examination of Graduate Education in Astronomy* (funded by the NSF; AAS, 1997). Activities at AAS meetings concerning employment, classroom education, and public policy serve to inform astronomers and help them in meeting their professional responsibilities.

The AAS has encouraged and participated in the broad community discussion that produced this decadal consensus on the future of astronomy. Public forums at the AAS scientific meetings were invaluable in communicating with a broad range of astronomers; the Web site supported by the AAS produced discussion and a variety of ideas. In addition, membership information from the AAS data collection system complemented other data used in this report.

The Astronomical Society of the Pacific (ASP) is an organization of both professional and amateur astronomers that has developed a strong outreach and education program for amateur astronomers and the public. It is the largest general astronomy society in the world, with members from more than 70 nations. The ASP's professional publications, a journal and a series of conference proceedings, are valuable resources for the astronomical community. Recently the ASP has focused on helping elementary school teachers use astronomy to excite children about careers in science, engineering, technology, and mathematics (see Chapter 5). The ASP's catalog (online at <www.aspsky.org/catalog.html>) is designed as an accessible resource to help teachers acquire lesson plans and demonstration materials for use in the classroom. The ASP also seeks to represent the astronomy instructors in 2-year colleges and junior colleges.

The American Physical Society's (APS's) Division of Astrophysics is dedicated to advancing and communicating knowledge of astrophysics and its relationship to the understanding of fundamental physical processes. It pays particular attention to linking astrophysics to nuclear and particle physics. The division convenes symposia in connection with APS meetings—for example, a symposium celebrating 100 years of astrophysics was held at the APS centennial meeting in 1999.

These professional societies also assist in identifying capable and visionary astronomers to meet the important challenges of working in short- and long-term positions at the federal agencies, particularly the NSF and NASA, where dedicated staff can have a significant effect on the progress of astronomy.

Action on several of the committee's recommendations depends on the future participation and leadership of the AAS, ASP, and APS. The committee urges that the funding agencies recognize and support specific activities of these broadly based professional organizations of astronomers.

CONGRESSIONAL QUESTIONS

The House of Representatives Committee on Science staff, citing 1997 authorization language for NSF, asked the NRC to respond to several questions, which were then divided between the Committee on Astronomy and Astrophysics (CAA) and the Astronomy and Astrophysics Survey Committee (AASC). The questions for which the CAA was responsible are addressed in the NRC report *Federal Funding of Astronomical Research* (NRC, 2000). The AASC's answers to its questions are implicit in the preceding discussions and recommendations and are presented explicitly here.

Have NASA and NSF mission objectives resulted in a balanced, broad-based, robust science program for astronomy? Have these overall missions been adequately coordinated and has this resulted in an optimum science program from a productivity standpoint?

Astronomy in the United States benefits from a vigorous program in space, coordinated by NASA, and a program of basic research and ground-based astronomy, led by the NSF. The DOE and other agencies contribute as well. The committee is generally pleased with the current and proposed ground- and space-based initiatives, which demonstrate that a robust and broadly based program is in place.

Balance among various components of the program, however, remains a concern of the astronomical community. A large portion of the total support for astronomy is now tied to a few flagship missions of NASA. The committee shares the concern, expressed in the report *Federal Funding of Astronomical Research* (NRC, 2000), that this arrangement leaves the program susceptible to a catastrophic mission failure. The committee's recommendation for a diverse range of missions addresses this issue to some extent.

To create a better balance between NASA and the NSF, the committee has made several recommendations to strengthen the ground-based program:

- National and independent observatories should be viewed as integrated systems—one for ground-based OIR astronomy, one for ground-based solar astronomy, and one for radio astronomy—of capabilities and resources for the United States as a whole. This approach leverages the private contributions to astronomy and positions the nation well for science opportunities in the international arena.
- Funds for grants should be included in the budgets of new ground-based facilities for their first 5 years of operation.
- The NSF should take more initiative in sharing the achievements of its scientists with the public, just as NASA does.
- The NSF should work with other agencies and with the astronomical community to build interagency programs that will aggressively pursue astronomical problems of broad national interest.

What special strategies are needed for strategic cooperation between NASA and NSF? Should these be included in agency strategic plans?

Coordination between NASA and the NSF can be advantageous because (1) the sheer scale of many modern astronomical investigations requires a coordinated national, if not international, effort; (2) the enormous increase in technical, computer, and Web-based capability reduces the barriers to cooperation; and (3) the growing sophistication of researchers' investigations implies that coordination across wavelength bands and across disciplines is required to produce a deeper and more fundamental understanding of the objects and processes under study. Interagency cooperation and joint projects between NASA and different divisions of NSF can initiate new scientific opportunities (e.g., GONG and SOHO-MDI in helioseismology) and capitalize on the NSF's many resources in research, engineering, and education. Successful collaborations between NASA and the NSF have focused on specific science issues in activities such as the Shoemaker-Levy Jupiter Collision and Life in Extreme Environments programs. All the scientific themes identified by the committee as being ripe for progress in this decade can be addressed by both ground- and space-based facilities, opening the way for additional NSF and NASA cooperation. Furthermore, the committee has recommended that the National Virtual Observatory and the National Astrophysical Theory Postdoctoral Program be jointly funded by NASA and the NSF.

In addition to NASA and the NSF, DOE sponsors research in astrophysics to address fundamental problems linked especially to particle physics, nuclear physics, and cosmology. Collaborations among these agencies are most effective when they are driven by specific scientific programs and when each agency contributes the special expertise of its area. Each agency can recognize its own unique capabilities and those of related agencies, and each should initiate the steps toward collaborations that it believes will be fruitful. Each agency should have a strategic plan for astronomy and astrophysics in place and should also have cross-disciplinary committees (such as DOE and NSF's Scientific Assessment Group for Experiments in Non-Accelerator Physics [SAGENAP] and NASA's Space Science Advisory Committee [SSAC]) available to evaluate major collaborative activities in astrophysics. The CAA should provide oversight from the NRC. The committee has recommended that these agencies should work together and with the astronomical community to build new interagency programs that will address astronomical problems of broad national interest. The traditional broker for interagency cooperation, the Office of Science and Technology Policy, could play a constructive role in facilitating the necessary coordination.

How do NASA and NSF determine the relative priority of new technological opportunities (including new facilities) compared to providing long-term support for associated research grants and facility operations?

At present NASA and the NSF differ in their approach to supporting new facilities, research grants, and facility operations. NASA is charged with developing scientific opportunities in space, and providing frequent access to space remains a paramount goal. With that framework, and with the aid of the scientific community, NASA has developed four scientific themes. Missions to develop these themes are budgeted with a total cost that includes construction, mission operations, and data analysis as one package in the mission's prime phase; the funding level for operations and data analysis during the mission's subsequent extended phase is determined by competitive review among all operating missions (NASA's Senior Review process). NSF AST supports facilities with the major share of its budget, and only recently has it developed a strategic plan. The University Grants program of the NSF supports investigator-initiated basic research that covers all of astronomy and astrophysics. When scientific opportunities are compelling, new facilities may be developed in ground-based astronomy. The NSF allots its construction monies through a major research equipment account line, distinct from the grants and operations accounts. Frequently, provision of funds to capitalize on the astronomy made possible by new facilities is neglected, preventing the new facilities from reaching their full potential and squeezing the NSF AST grants program. In both cases basic research suffers.

The committee believes that the NSF astronomy program would be strengthened if the budgeting and operations procedure were changed to include adequate funds for operations, instrumentation, and grants associated with each new facility. In addition the committee proposes cross-disciplinary competitive reviews of major ground-based facilities to evaluate the allocation of resources, with the aim of optimizing the scientific return on the nation's investment in astronomy and astrophysics.

NOTES

1. The 8-m-class Japanese telescope, Subaru, has just been completed on Mauna Kea. The European Southern Observatory's Very Large Telescope—a grouping of four 8-m-class telescopes in Chile that can be used individually or linked together—is now under construction, with two of the four now in operation. Also entering construction is the Gran Telescopio Canarias (10 m+) of the Institute for Astrophysics in the Canary Islands.

2. Members of the group included Frank Bash, chair, Don Campbell, Bruce Carney, Richard Elston, Phil Goode, Ken Kellermann, David Morrison, Thomas Rimmele, Blair Savage, and Stephen Strom.

3. Evaluating the percentage of a country's gross domestic product (GDP) per citizen that is invested in 8-m-class telescopes, L. Ramsey (personal communication, 1999) found that this number amounted to 0.0157 percent for European Southern Observatory-member countries and 0.0125 percent for Japan, but only 0.0011 percent for the United States—a factor of 10 lower for the United States. Indeed, compared with the contributions of all the member countries participating in the U.S. national telescope, Gemini, the U.S. contribution ranked next to last in percentage of GDP invested per citizen; only Brazil ranked lower.

4. The Owens Valley Radio Observatory and the Caltech Submillimeter Observatory give 50 percent of their time to outside observers; the Berkeley-Illinois-Maryland Association allocates 30 percent of observing time to outside astronomers; the Five College Radio Astronomy Observa-

tory awards 50 percent of time on the 14-m millimeter-wave telescope to external users (observatory directors, personal communications, 2000).

5. "A Longitudinal Study of Selected Astronomers Based on Early Sources of Support," provided by D. Helfand, Columbia University, to the Panel on Astronomy Education and Policy.

6. By contrast, Asians represented 3 percent of the U.S. population but constituted 10 percent of the scientists and engineers in the United States in 1995 (NSB, 1996).

7. Data in a personal communication from K. Marvel, AAS, 1999.

8. NSF tabulation. Statistics on minorities in astronomy faculty positions are not currently available. Data from an AAS-initiated sequence of frequently conducted member surveys should become available in the future.

References

American Astronomical Society (AAS). 1997. *The American Astronomical Society's Examination of Graduate Education in Astronomy,* AAS Education Board and AAS Graduate Advisory Board. *Bulletin of the American Astronomical Society,* vol. 29, no. 5.

American Astronomical Society Committee on the Status of Women in Astronomy (AAS CSWA). 1996. *STATUS: A Report on Women in Astronomy,* June, pp. 2-4, J.A. White. AAS, Washington, D.C.

American Astronomical Society Committee on the Status of Women in Astronomy (AAS CSWA). 2000. *STATUS: A Report on Women in Astronomy,* June, pp. 1-4, 7, M. Urry. AAS, Washington, D.C.

American Institute of Physics (AIP). 2000. *Enrollments and Degrees Report,* AIP R-151.35, P.J. Mulvey and S. Nicholson. AIP, College Park, Md.

European Southern Observatory (ESO). 1998. *El Mensajero, ESO Newsletter,* no. 91, March, Table 1.

Executive Office of the President, Office of Science and Technology Policy. 1994. *Science in the National Interest,* President William J. Clinton and Vice President Albert Gore, Jr. Government Printing Office, Washington, D.C.

Massachusetts Institute of Technology (MIT). 1999. "A Study on the Status of Women Faculty in Science at MIT," *The MIT Faculty Newsletter,* vol. XI, no. 4, pp. 4-13 [summary of a 150-page unpublished report on the same subject prepared in 1994]. MIT, Cambridge, Mass.

National Center for Education Statistics (NCES), U.S. Department of Education. 1996. *1996 Mathematics Report Card for the Nation and the States: Findings from the National Assessment of Educational Progress,* C.M. Reese, K.E. Miller, J. Mazzeo, and J.A. Dossey. GPO No. 065-000-00984-6. Government Printing Office, Washington, D.C.

National Research Council (NRC). 1964. *Ground-based Astronomy: A Ten-Year Program.* National Academy of Sciences, Washington, D.C.

National Research Council (NRC). 1972. *Astronomy and Astrophysics for the 1970's.* National Academy of Sciences, Washington, D.C.

National Research Council (NRC). 1982. *Astronomy and Astrophysics for the 1980's. Volume I: Report of the Astronomy Survey Committee.* National Academy Press, Washington, D.C.

National Research Council (NRC). 1991. *The Decade of Discovery in Astronomy and Astrophysics.* National Academy Press, Washington, D.C.

National Research Council (NRC). 1995. *A Strategy for Ground-Based Optical and Infrared Astronomy.* National Academy Press, Washington, D.C.

National Research Council (NRC). 1996. *National Science Education Standards.* National Academy Press, Washington, D.C.

National Research Council (NRC). 1997. *A New Science Strategy for Space Astronomy and Astrophysics.* National Academy Press, Washington, D.C.

National Research Council (NRC). 1999. *Gravitational Physics: Exploring the Structure of Space and Time.* National Academy Press, Washington, D.C.

National Research Council (NRC). 2000. *Federal Funding of Astronomical Research.* National Academy Press, Washington, D.C.

National Research Council (NRC). 2001. *Astronomy and Astrophysics in the New Millennium: Panel Reports.* National Academy Press, Washington, D.C.

National Research Council and European Science Foundation (NRC-ESF). 1999. *U.S.-European Collaboration in Space Science.* National Academy Press, Washington, D.C.

National Science Board (NSB). 1996. *Science and Engineering Indicators—1996,* NSB 96-21. Government Printing Office, Washington, D.C.

National Science Foundation (NSF). 1994. *Women, Minorities, and Persons with Disabilities in Science and Engineering: 1994,* NSF 94-333, Division of Science Resources Studies. NSF, Arlington, Va.

National Science Foundation (NSF). 1996. *Selected Data on Science and Engineering Doctorate Awards 1995,* NSF 96-303, S.T. Hill, project officer, Division of Science Resources Studies. NSF, Arlington, Va.

National Science Foundation (NSF). 1999a. *Science and Engineering Doctorate Awards: 1997,* NSF 94-333, Division of Science Resources Studies. NSF, Arlington, Va.

National Science Foundation (NSF). 1999b. "Doctorate Awards Declining in Some Science and Engineering Fields," Data Brief, April 6, NSF 99-339, S.T. Hill, Division of Science Resources Studies. NSF, Arlington, Va.

Sonnert, G., and Holton, G. 1996. "Career Patterns of Women and Men in the Sciences," *American Scientist,* vol. 84, pp. 63-71.

Valian, V. 1998. *Why So Slow? The Advancement of Women.* MIT Press, Cambridge, Mass.

Appendix—Definitions

ASTRONOMICAL TERMS

Accretion, accretion disk—Astronomical objects as diverse as *protostars* and *active galaxies* may derive their energy from the gravitational power released by the infall, or accretion, of material onto a central object. The combined effects of gravity and rotation often force the accreting material into an orbiting accretion disk.

Active galaxy—Certain galaxies emit far more energy than can be accounted for by their stars alone. The central regions of these galaxies harbor a compact, solar-system-sized object capable of outshining the rest of the galaxy by a factor of 100. The ultimate energy source for active galaxies may be the *accretion* of matter onto a *supermassive black hole*. Active galaxies can emit strongly across the entire *electromagnetic spectrum*, from radio waves to gamma rays. See *quasar*.

Active optics—A technique to reduce the effects of slowly varying forces, such as gravitational deflections and temperature drifts, that can distort a mirror on time scales of minutes to hours, resulting in imperfect images.

Adaptive optics—A set of techniques to adjust the mirrors of telescopes on time scales of hundredths of a second to correct for distortions in astronomical images due to turbulence in Earth's atmosphere.

Anisotropy—Dependence of the properties of a system on the orientation or the direction of observation. The distribution of galaxies in space is not uniform, whereas the intensity of the *cosmic background radiation* from the *Big Bang* is highly uniform in all directions—i.e., it is almost isotropic. Astronomers are using sensitive telescopes to study the small anisotropies in the cosmic background radiation that should be present given the non-uniform distribution of galaxies.

Arcminute—A unit of angle corresponding to 1/60th of a degree. The full moon is 30 arcminutes in diameter.

Arcsecond—A unit of angle corresponding to 1/3600th of a degree; 1/60th of an *arcminute*. An arcsecond is approximately the size of a dime viewed from a distance of 1 mile.

Array—There are two examples of arrays in common use in astronomy: (1) A group, or array, of telescopes can be combined to simulate a single large telescope, kilometers or even thousands of kilometers across.

(2) Astronomical instruments have recently been fabricated using new electronic components called detector arrays or *charge-coupled devices* (CCDs) that consist of thousands of individual detectors constructed on centimeter-sized wafers of silicon, or other materials.

Astrometry—The branch of astronomy concerned with measuring the positions of celestial objects. Advances in technology may soon permit a 1,000-fold improvement in the measurement of positions, and thus in astronomers' ability to determine distances to stars and galaxies. See *parallax*.

Bahcall report—The National Research Council decadal survey report, *The Decade of Discovery in Astronomy and Astrophysics* (1991), on astronomy and astrophysics for the 1990s, chaired by J.N. Bahcall.

Baseline—The separation between telescopes in an *interferometer*. The largest baseline determines the finest detail that can be discerned with an interferometer.

Big Bang—Most astronomers believe that the universe began in a giant explosion called the Big Bang about 14 billion years ago. Starting from an initial state of extremely high density, the universe has been expanding and cooling ever since. Some of the most fundamental observed properties of the universe, including the abundance of light elements such as helium and lithium and the recession of galaxies, can be accounted for by modern theories of the Big Bang.

Black hole—A region in space where the density of matter is so extreme, and the resultant pull of gravity so strong, that not even light can escape. Black holes are probably the end point in the evolution of some types of stars and are probably located at the centers of some *active galaxies* and *quasars*.

Blackbody radiation—A glowing object emits radiation in a quantity and at wavelengths that depend on the temperature of the object. For example, a poker placed in a hot fire first glows red-hot, then yellow-hot, then finally white-hot. This radiation is called thermal or blackbody radiation.

Brown dwarf—A star-like object that contains less than about 0.08 the mass of the Sun and is thus too small to ignite nuclear fuels and become a normal star. Brown dwarfs emit small amounts of infrared radiation due to the slow release of gravitational energy and may be a component of *dark matter*.

Byte—A unit of information used in reference to computers and quantities of data. A byte consists of 8 bits (0's and 1's) and may correspond to a single character or number.

Carbon monoxide—A molecule (CO) consisting of carbon and oxygen that emits strongly at millimeter and *submillimeter* wavelengths and that can be used to trace cool gas in our own and other *galaxies*.

Charge-coupled device, or CCD—An electronic detector used for low-light-level imaging and astronomical observations. CCDs were developed by NASA for use in the Hubble Space Telescope and the Galileo probe to Jupiter and are now widely used on ground-based telescopes. See also *array*.

Cosmic microwave background radiation—The radiation left over from the *Big Bang* explosion at the beginning of the universe. As the universe expanded, the temperature of the fireball cooled to its present level of 2.7 degrees above absolute zero (2.7 K). *Blackbody radiation* from the cosmic background is observed at radio, millimeter, and *submillimeter* wavelengths.

Cosmic rays—Protons and nuclei of heavy atoms that are accelerated to high energies in the magnetic field of our galaxy and that can be studied directly from Earth or from satellites.

Dark energy—An as yet unknown form of energy that pervades the universe. Its presence was inferred from the discovery that the expansion of the universe is accelerating, and these observations suggest that about 70 percent of the total density of matter plus energy is in this form. Such an acceleration would be predicted if the cosmological constant that Einstein included in his general theory of relativity were non-zero.

Dark matter—Approximately 80 percent of the matter in the universe may so far have escaped direct detection. The presence of this unseen matter has been inferred from motions of stars and gas in *galaxies*, and of galaxies in clusters of galaxies. Candidates for the missing mass include *brown dwarf* stars and exotic subatomic particles. Dark matter was called "missing mass" for many years. However, because it is the light, not the mass, that is missing, astronomers have given up this terminology.

Diffraction limit—The finest detail that can be discerned with a telescope. The physical principle of diffraction limits this level of detail to a value proportional to the wavelength of the light observed divided by the diameter of the telescope.

Electromagnetic spectrum—Radiation can be represented as electric and magnetic fields vibrating with a characteristic wavelength or frequency. Long wavelengths (low frequencies) correspond to radio radiation; intermediate wavelengths, to millimeter and infrared radiation; short wavelengths (high frequencies), to visible and ultraviolet light; and extremely short wavelengths, to x rays and gamma rays. Most astronomical observations measure some form of electromagnetic radiation.

Expansion of the universe—The tendency of every part of the universe to move away from every other part due to the initial impetus of the *Big Bang;* also known as the Hubble expansion, after the American astronomer Edwin Hubble, whose observations of receding *galaxies* led to scientists' current understanding of the expanding universe. See *redshift.*

Extragalactic—Objects outside our *galaxy*, more than about 50,000 light-years away, are referred to as extragalactic.

Fly's Eye—A cosmic-ray telescope used to monitor gamma rays from astronomical sources.

Galaxy—An isolated grouping of tens to hundreds of billions of stars ranging in size from 5,000 to 150,000 light-years across. Spiral galaxies like our own *Milky Way* are flattened disks of stars and often contain large amounts of gas out of which new stars can form. Elliptical galaxies are shaped more like footballs and are usually devoid of significant quantities of gas.

Gamma-ray astronomy—The study of astronomical objects using the most energetic form of electromagnetic radiation.

General relativity—Einstein's theory of gravity in which the gravity is the curved geometry of space and time.

Gigabyte—One billion (10^9) *bytes*. A unit of information used to describe quantities of data or the storage capacity of computers.

Gravitational lens—A consequence of Einstein's general relativity theory is that the path of light rays can be bent by the presence of matter. Astronomers have observed that the light from a distant galaxy or quasar can be "lensed" by the matter in an intervening galaxy to form multiple and often distorted images of the background object.

Great Observatories—A NASA program to launch four major observatories to cover the optical (HST), gamma-ray (CGRO), x-ray (Chandra), and infrared (SIRTF) portions of the electromagnetic spectrum.

Halo (of a galaxy)—The roughly spherical distribution of dark matter and thinly scattered stars, star clusters, and gas that surround a spiral galaxy.

Helioseismology—The study of the internal vibrations of the Sun. In a manner analogous to terrestrial seismology, helioseismology can reveal important information about the Sun's internal condition.

Hubble Space Telescope (HST)—A 2.4-m-diameter space telescope designed to study visible, ultraviolet, and infrared radiation; the first of NASA's Great Observatories.

Hydrogen—The most abundant element in the universe. It can be observed at a variety of wavelengths, including 21-cm radio, infrared, visible, and ultraviolet wavelengths, and in a variety of forms, including atoms (HI) and molecular (H_2) and ionized (HII) forms.

Infrared astronomy—The study of astronomical objects using intermediate-wavelength radiation to which the atmosphere is mostly opaque and the human eye insensitive. Humans sense infrared energy as heat. The infrared part of the *electromagnetic spectrum* generally corresponds to radiation with wavelengths from 1 μm to 1,000 μm. Objects with temperatures around room temperature or lower emit most of their radiation in the infrared.

Interferometer, interferometry—A spatial interferometer combines beams of light from different telescopes to synthesize the aperture of a single large telescope; see *array*. Spatial interferometry is the main technique used by astronomers to map sources at high resolution and to measure their positions with high precision. A different form of interferometer can be used on a single telescope to break up the light into its constituent colors; see *spectroscopy*.

Light-year—A unit of astronomical distance equal to the distance light travels in a year: about 9 trillion miles. The nearest star is 4 light-years away. The center of our galaxy is about 25,000 light-years away. The closest galaxy is about 180,000 light-years away.

Magellanic Clouds, Large and Small—The two closest galaxies to our own Milky Way, located about 180,000 light-years away and visible only from the Southern Hemisphere. A bright *supernova*, SN1987A, was observed in the Large Magellanic Cloud in 1987.

Magnetohydrodynamics—The study of the motion of gases in the presence of magnetic fields.

Magnitude—A unit of brightness for stars. Fainter stars have numerically larger magnitudes. The brightest stars, excluding the Sun, are about magnitude 0; the faintest star visible to the unaided eye is about magnitude 6. A star of magnitude 15 is one-millionth as bright as the half-dozen brightest stars of magnitude 0. Stars as faint as magnitude 28 can be seen with powerful terrestrial or spaceborne telescopes.

Massive compact halo object (MACHO)—An object of roughly stellar mass in the halo of our galaxy that is too faint to be detected by its own emission. MACHOs are indirectly detected via gravitational microlensing of more distant stars.

Megabyte—One million *bytes*. A unit of information used to describe quantities of data or the storage capacity of computers. A single image from the Hubble Space Telescope comprises about 5 megabytes.

Microlensing—Gravitational lensing due to a stellar mass object. This lensing phenomenon is called microlensing because the mass of the lens is so small compared with that of a galaxy. Microlensing of distant stars by intervening faint stars can reveal planets in orbit around the lensing star.

Milky Way—Our Sun is located in the Milky Way Galaxy, a spiral galaxy consisting of some 100 billion stars spread in a disk more than 80,000 light-years across and hundreds of light-years thick. The central disk of the Milky Way is the wide path of faint light that stretches across the night sky.

Neutrino—One of a family of subatomic particles with little or no mass. These particles are generated in nuclear reactions on Earth, in the centers of stars, and during *supernova* explosions and can give unique information about these energetic processes. Because neutrinos interact only weakly with matter, they are difficult to detect.

Nucleosynthesis—The process by which heavy elements such as helium, carbon, nitrogen, and iron are formed out of the fusion of lighter elements, such as hydrogen, during the normal evolution of stars, during *supernova* explosions, and in the *Big Bang*.

Optical astronomy—The study of astronomical objects using light waves with wavelengths from about 1 to 0.3 μm. The human eye is sensitive to most of these wavelengths. See *electromagnetic spectrum*.

Parallax—The apparent shift in position of a nearby object relative to a more distant object, as the observer changes position. Using basic trigonometry, it is possible to derive the distance of a star from its parallax as observed from opposite points on Earth's orbit. See *astrometry*.

Parsec—A unit of astronomical distance equal to 3.2616 light-years.

Pixel—The smallest element of a digital image. A typical image from the *Hubble Space Telescope* is a square with $1{,}600 \times 1{,}600$ discrete pixels.

Protogalaxy—*Galaxies* are thought to have formed fairly early in the history of the universe, by the collapse of giant clouds of gas. During this process, a first generation of stars formed, and these should be observable with the telescopes discussed in this report.

Protoplanetary or protostellar disk—A disk of gas and dust surrounding a young star or *protostar* out of which planets may form.

Protostar—The earliest phase in the evolution of a star, in which most of its energy comes from the infall of material, or *accretion*, onto the growing star. A *protostellar disk* probably forms around the star at this time.

Quasar—An extremely compact, luminous source of energy found in the cores of certain galaxies. A quasar may outshine its parent galaxy by a factor of 1,000 yet be no larger than our own solar system. The *accretion* of gas onto a *supermassive black hole* probably powers the quasar. *Active galaxies* are probably less luminous and less distant versions of quasars.

Radio astronomy—The study of astronomical objects using radio waves with wavelengths generally longer than 0.5 to 1 mm. See *electromagnetic spectrum*.

Redshift—Radiation from an approaching object is shifted to higher frequencies (to the blue), while radiation from a receding object is shifted to lower frequencies (to the red). A similar effect raises the pitch of an ambulance siren as it approaches. The *expansion of the universe* makes objects recede so that the light from distant galaxies is redshifted. The redshift is parameterized by z, where the wavelength shift is given by the factor $(1 + z)$ times the wavelength.

Resolution—Spatial resolution describes the ability of an instrument to separate features at small details; see *diffraction limit* and *interferometer*. Spectral resolution describes the ability of an instrument to discern small shifts in wavelength; see *spectroscopy*.

Spectroscopy—A technique whereby the light from astronomical objects is broken up into its constituent colors. Radiation from the different chemical elements that make up an object can be distinguished, giving information about the abundances of these elements and their physical state.

Starburst galaxy—Certain *galaxies*, particularly those perturbed by a close encounter or collision with another galaxy, often form stars at a rate hundreds of times greater than that evident in our galaxy. Such galaxies are bright sources of *infrared* radiation.

Submillimeter radiation—*Electromagnetic* radiation with wavelengths between about 0.1 and 1 mm intermediate between radio and infrared radiation.

Sunyaev-Zeldovich effect—An astrophysical effect whereby the distribution of wavelengths of radiation seen through the gas in a distant cluster of galaxies is subtly modified. Measurement of this effect can be used to determine the distance to the cluster.

Supermassive black hole—A *black hole* that is much more massive than the Sun. Supermassive black holes with masses exceeding a million solar masses are found in the nuclei of most *galaxies*.

Supernova—A star that, due to *accretion* of matter from a companion star or exhaustion of its own fuel supply, can no longer support itself against its own weight and thus collapses, throwing off its outer layers in a burst of energy that outshines an entire galaxy. In 1987 a star in the *Large Magellanic Cloud* was observed as a dramatic supernova called Supernova 1987A.

Terabyte—One trillion (10^{12}) *bytes*. A unit of information used to measure quantities of data. All the images taken with the Hubble Space Telescope in a given year will comprise a few terabytes.

Triad—An experimental low-altitude spacecraft launched in 1972 for gravitational physics tests.

Ultraviolet (UV) astronomy—The study of astronomical objects using short-wavelength radiation, from 0.3 μm to 0.01 μm (10 nm), to which the atmosphere is opaque and the human eye insensitive. See *electromagnetic spectrum*.

X-ray astronomy—The study of astronomical objects using x rays with wavelengths shorter than about 10 nm, to which the atmosphere is opaque. X rays are emitted by extremely energetic objects that have temperatures of millions of degrees. See *electromagnetic spectrum*.

z—See *redshift*.

ABBREVIATIONS AND ACRONYMS

ACCESS—The Advanced Cosmic-ray Composition Experiment for the Space Station, a cosmic-ray experiment on the International Space Station.

AIPS—Astronomical Image Processing System. A set of programs developed to process astronomical data from the Very Large Array (VLA) and other radio wavelength *interferometers*.

ALMA—The Atacama Large Millimeter Array.

AO—See *adaptive optics*.

ARISE—The Advanced Radio Interferometry between Space and Earth, an orbiting antenna that will combine with the ground-based VLBA.

ASCA—The Japanese Advanced Satellite for Cosmology and Astrophysics mission.

AST—The Advanced Solar Telescope; also, the NSF's Division of Astronomical Sciences.

ATCA—Australia Telescope Compact Array. An interferometric radio array consisting of six antennae.

ATNF—Australia Telescope National Facility. The main national organization supporting and undertaking research in radio astronomy.

AU—Astronomical unit. A basic unit of distance equal to the separation between Earth and the Sun, about 150 million km.

AXAF—The Advanced X-ray Astrophysics Facility; now called Chandra, after astrophysicist S. Chandrasekhar.

BATSE—The Burst and Transient Spectrometer Experiment aboard CGRO, an all-sky gamma-ray-burst monitor.

BIMA—Berkeley-Illinois-Maryland Association. A consortium that operates a millimeter-wave radio interferometer at Hat Creek, California.

BOOMERANG—Balloon Observations of Millimetric Extragalactic Radiation and Geophysics, an extremely sensitive microwave telescope flown under a stratospheric balloon that circumnavigated Antarctica over 11 days in late 1998 and early 1999.

CARMA—The Combined Array for Research in Millimeter-wave Astronomy, a millimeter-wave array in the Northern Hemisphere.
CCD—See *charge-coupled dcvice.*
CGRO—The Compton Gamma Ray Observatory. A telescope launched in 1991 to study highly energetic gamma rays from astronomical sources. NASA's second Great Observatory. It was safely deorbited and reentered Earth's atmosphere in June 2000.
Chandra—An x-ray observatory launched in 1999. NASA's third Great Observatory.
CMB—See *cosmic microwave background radiation.*
CO—See *carbon monoxide.*
COBE—The Cosmic Background Explorer. A NASA mission launched in 1989 to study the *cosmic background radiation* from the *Big Bang*.
Con-X—The Constellation X-ray Observatory, a suite of four powerful x-ray telescopes.
CSO—The Caltech Submillimeter Observatory, a 10-m telescope operating on Mauna Kea, Hawaii. The telescope is used for observations of millimeter and *submillimeter* wavelength radiation.

DOD—Department of Defense.
DOE—Department of Energy.

EGRET—The Energetic Gamma Ray Experiment aboard the Compton Gamma Ray Observatory.
ESA—European Space Agency. The European equivalent of NASA.
ESO—The European Southern Observatory.
EUVE—The Extreme Ultraviolet Explorer NASA mission.
EVLA—The Expanded Very Large Array.
EXIST—The Energetic X-ray Imaging Survey Telescope, which will be attached to the ISS.

FASR—The Frequency Agile Solar Radio telescope.
FCRAO—The Five College Radio Astronomy Observatory.
FIRST—The European Far Infrared Space Telescope.

GBT—The Green Bank Telescope.
GLAST—The Gamma-ray Large Area Space Telescope, a joint NASA-DOE mission.
GMRT—The Giant Metrewave Radio Telescope in India.

GONG—Global Oscillations Network Group. A worldwide network of telescopes designed to study vibrations in the Sun. See *helioseismology*.
GSFC—Goddard Space Flight Center.
GSMT—The Giant Segmented Mirror Telescope, a 30-m-class, ground-based telescope.

HALCA—An 8-m radio telescope satellite and the key component of Japan's VLBI Space Observatory Program.
HEAO—The High-Energy Astronomical Observatory. A series of three telescopes launched in the late 1970s and early 1980s to study x rays and gamma rays. HEAO-2 was also called the Einstein Observatory and was the first fully imaging x-ray telescope put into space.
HESSI—The High Energy Solar Spectroscopic Imager.
HET—The Hobby-Eberly Telescope.
HETE-2—The High-Energy Transient Explorer, a small mission launched to look for the origins of mysterious bursts of x rays and gamma rays.
HI, HII, H2—Respectively, atomic hydrogen, ionized hydrogen, and molecular hydrogen.
HST—The *Hubble Space Telescope.*

IRAF—Image Reduction and Analysis Facility. A set of computer programs for working with astronomical images.
IRAM—Institut de Radioastronomie Millimétrique. An international (French, German, and Spanish) institute for research in millimeter astronomy.
IRAS—The Infrared Astronomical Satellite. A NASA Explorer satellite launched in 1983 that surveyed the entire sky in four infrared wavelength bands using a helium-cooled telescope.
ISS—International Space Station.
IUE—The International Ultraviolet Explorer. A joint NASA-ESA orbiting telescope to study ultraviolet radiation.

JCMT—The James Clerk Maxwell Telescope.

LBT—The Large Binocular Telescope.
LHC—The Large Hadron Collider. A particle accelerator under construction at the European Laboratory for Particle Physics.
LIGO—The Laser Interferometer Gravitational-wave Observatory, an NSF-sponsored project to build and operate two 4-km laser interferometers to detect gravitational waves.

LISA—The Laser Interferometer Space Antenna.
LMT—The Large Millimeter Telescope.
LOFAR—The Low Frequency Array, a joint Dutch-U.S. initiative to study radio wavelengths longer than 2 m.
LSST—The Large-aperture Synoptic Survey Telescope, a 6.5-m-class optical telescope.
LTSA—Long-Term Space Astrophysics research program sponsored by NASA's Office of Space Science.

MACHO—See *massive compact halo object.*
MAP—The Microwave Anisotropy Probe mission.
MAXIMA—Millimeter Anisotropy experiment Imaging Array, a balloon-borne millimeter-wave telescope designed to measure the angular power spectrum of fluctuations in the CMB over a wide range of angular scales.
MERLIN—The Multi-Element Radio Linked Interferometer Network, an array of radio telescopes distributed around Great Britain and operated by the Jodrell Bank Observatory.
MIDEX—Mid-size Explorer mission.
MMA—The Millimeter Array, now part of ALMA.
MMT—The Multiple Mirror Telescope.
MO&DA—NASA Mission Operations and Data Analysis.

NAIC—National Astronomy and Ionosphere Center.
NAS—National Academy of Sciences.
NASA—National Aeronautics and Space Administration.
NGST—The Next Generation Space Telescope, an 8-m-class infrared space telescope.
NOAO—The National Optical Astronomy Observatories.
NRAO—The National Radio Astronomy Observatory.
NRC—National Research Council.
NSF—National Science Foundation.
NSO—The National Solar Observatory.
NVO—The National Virtual Observatory, a "virtual sky" based on enormous data sets.

1HT—One Hectare Telescope.
OSS—NASA's Office of Space Science.
OVRO—Owens Valley Radio Observatory.

Planck Surveyor—An ESA-led space mission to image anisotropies in the CMB.

RHIC—The DOE's Relativistic Heavy Ion Collider at Brookhaven National Laboratory.
ROSAT—The Roentgen Satellite, an orbiting x-ray telescope launched in 1990, is named after the German scientist W. Röntgen, the discoverer of x-rays. ROSAT is a German-U.S.-U.K. collaboration.
ROTSE—Robotic Optical Transient Search Experiment.
RXTE—The Rossi X-ray Timing Explorer, a NASA mission.

SAFIR—The Single Aperture Far Infrared Observatory, an 8-m-class space-based telescope.
SAGENAP—The DOE and NSF's Scientific Assessment Group for Experiments in Non-Accelerator Physics.
SALT—The Southern African Large Telescope.
SDO—The Solar Dynamics Observer, a successor to the pathbreaking SOHO mission.
SETI—Search for extraterrestrial intelligence.
SIM—The Space Interferometry Mission.
SIRTF—The Space Infrared Telescope Facility. NASA's fourth Great Observatory will study infrared radiation.
SKA—The Square Kilometer Array, an international centimeter-wave radio telescope.
SMA—The Submillimeter Array.
SMEX—Small Explorer. A NASA program to fly small, inexpensive satellites on a rapid timetable.
SOFIA—The Stratospheric Observatory for Infrared Astronomy.
A 2.5-m telescope flown above most of Earth's water vapor in a modified 747 aircraft to study *infrared* and *submillimeter* radiation.
SOHO—The joint ESA-NASA Solar and Heliospheric Observatory.
SOHO-MDI—The SOHO-Michelson Doppler Imager.
SOLIS—The Synoptic Optical Long-term Investigation of the Sun, a ground-based facility for solar observations that is funded by the NSF and designed and built by NSO with planned operations beginning in 2001.
SPST—The South Pole Submillimeter-wave Telescope.
SSAC—NASA's Space Science Advisory Committee.
STEREO—The Solar-Terrestrial Relations Observatory.
STScI—Space Telescope Science Institute.

TPF—The Terrestrial Planet Finder, designed to study terrestrial planets around nearby stars. It is currently envisaged as a free-flying infrared interferometer

TRACE—Transition Region and Coronal Explorer, a current NASA mission.

TSIP—The Telescope System Instrumentation Program.

2MASS—Two Micron All Sky Survey, a current ground-based project funded by the NSF and NASA.

ULDB—NASA's Ultralong-Duration Balloon program.

VERITAS—The Very Energetic Radiation Imaging Telescope Array System, designed to study energetic gamma rays using ground-based telescopes.

VLA—The Very Large Array. A radio interferometer consisting of 27 antennas spread out over 35 km and operating with 0.1-*arcsecond resolution.*

VLBA—The Very Long Baseline Array. An array of radio telescopes operating as an interferometer with a transcontinental *baseline* and *resolution* of less than a thousandth of an *arcsecond.*

VLBI—Very long baseline interferometry. A technique whereby a network of radio telescopes can operate as an *interferometer* with *baselines* that can be as large as the diameter of Earth, or even larger when satellites are used.

VLT—The Very Large Telescope. The European Southern Observatory's four 8-m telescopes.

VLT ISAAC—VLT Infrared Spectrometer and Array Camera.

WFC3—Wide Field Camera 3, an imaging instrument planned to replace the existing imaging camera onboard HST.

XMM-Newton—European x-ray space mission launched in 1999.

Index

A

B

C

D

E

F

G

H

I

N

P

Q

R

S

T

U

V

W

X

Y